Award Winning
Green Roof Designs

Green Roofs
for Healthy Cities

Steven W. Peck, Founder and President, Green Roofs for Healthy Cities, Honorary ASLA

Foreword by Arthur Charles Erickson, CCBArch, FRAIC, FAIA (hon), FRAIS (hon) and Cornelia Han Oberlander, C.M., FCSLA, FASLA

4880 Lower Valley Road, Atglen, PA 19310

Printed on 100%
post consumer
recycled paper

Schiffer Books are available at special discounts for bulk purchases for sales promotions or premiums. Special editions, including personalized covers, corporate imprints, and excerpts can be created in large quantities for special needs. For more information contact the publisher:

Published by Schiffer Publishing Ltd.
4880 Lower Valley Road
Atglen, PA 19310
Phone: (610) 593-1777; Fax: (610) 593-2002
E-mail: Info@schifferbooks.com

For the largest selection of fine reference books on this and related subjects, please visit our web site at **www.schifferbooks.com**
We are always looking for people to write books on new and related subjects. If you have an idea for a book please contact us at the above address.

This book may be purchased from the publisher.
Include $3.95 for shipping.
Please try your bookstore first.
You may write for a free catalog.

In Europe, Schiffer books are distributed by
Bushwood Books
6 Marksbury Ave.
Kew Gardens
Surrey TW9 4JF England
Phone: 44 (0) 20 8392-8585; Fax: 44 (0) 20 8392-9876
E-mail: info@bushwoodbooks.co.uk
Website: www.bushwoodbooks.co.uk
Free postage in the U.K., Europe; air mail at cost.

Copyright © 2008 by Steven W. Peck
Library of Congress Control Number: 2001012345

All rights reserved. No part of this work may be reproduced or used in any form or by any means—graphic, electronic, or mechanical, including photocopying or information storage and retrieval systems—without written permission from the publisher.
The scanning, uploading and distribution of this book or any part thereof via the Internet or via any other means without the permission of the publisher is illegal and punishable by law. Please purchase only authorized editions and do not participate in or encourage the electronic piracy of copyrighted materials.
"Schiffer," "Schiffer Publishing Ltd. & Design," and the "Design of pen and ink well" are registered trademarks of Schiffer Publishing Ltd.

Designed by rOs
Type set in Bernhard Modern BT/Arrus BT

ISBN: 978-0-7643-3022-3
Printed in China

Dedication

To Emily and Oliver and everyone with the imagination to dream of a better world, and the courage to work towards realizing it.

> "Hold fast to dreams
> For if dreams die
> Life is a broken-winged bird
> That cannot fly.
> Hold fast to dreams
> For when dreams go
> Life is a barren field
> Frozen with snow."
>
> — Langston Hughes (1902-1967)

Green Roofs for Healthy Cities – North America, Inc. was founded in 1999 as a small network of public and private organizations and is now a rapidly growing 501(c)(6), not-for-profit industry association. Our mission is to increase the awareness of the economic, social, and environmental benefits of green roofs and green walls, and other forms of living architecture through education, advocacy, professional development, and celebrations of excellence. See Appendix A for more information or visit www.greenroofs.org.

The Green Roof Awards of Excellence was launched in 2002 to identify green roof projects that demonstrate extraordinary vision and leadership, centered around the integration of green roofs with building systems, the building occupants, the surrounding site and larger community. The Awards of Excellence case studies in this book recognize and celebrate the valuable contributions of green roof owners, designers and installation professionals that have stretched the boundaries of green roof form and function. The Awards also serve to increase general awareness of green roofs, green walls and other forms of living architecture while promoting their associated public and private benefits.

Acknowledgments

This book would not have been possible without the unwavering dedication and professionalism of Flavia Bertram, who interviewed all of the award winners and updated project profiles and photos for the manuscript. Special thanks to Caroline Nolan and Joyce McLean, who provided invaluable editorial commentary throughout the process and to Jennifer Sprout and Hazel Farley, without whose ongoing support and dedication, Green Roofs for Healthy Cities would only be a shadow of its current self. Enormous gratitude and thanks to Chris M. Peck, who has been a source of support and inspiration for Green Roofs for Healthy Cities from the very beginning. Dr. Brad Bass is to be commended for his vision and leadership as well as the many researchers who continue to make important and lasting contributions to the practice of living architecture.

Thanks are also extended to the owners, designers, and manufacturers who contributed to their project profiles for this book, and to the many who have submitted their projects for Awards over the years. Very special thanks go to all the professionals who have invested their time and energy in judging the Green Roof Awards of Excellence over the past five years:

Paul Adelmann, Manager, Community & Local Government Relations
Jeffrey Bruce, President, Jeffrey L. Bruce & Company
Monica Kuhn, Architect
Paul Farmer, President, American Planning Association
Michael Gibbons, President, Architectural Systems Inc.
Robert Herman, Horticultural Consultant, Uncommon Plants
N. Marcia Jiménez, Commissioner, City of Chicago Department of the Environment
Tom Liptan, Landscape Architect, Environment Specialist, City of Portland
Steven Peck, Founder and President, Green Roofs for Healthy Cities
Ed Snodgrass, President, Emory Knoll Farms
Nancy Somerville, Executive Vice President, American Society of Landscape Architects
Bill Thompson, Editor, *Landscape Architecture Magazine*

We would also like to recognize the original board members of Green Roofs for Healthy Cities - North America Inc. for their immense contribution to the development of the organization.

Jeffrey L. Bruce, Jeffrey L Bruce & Co.
Leslie Hoffman, Earth Pledge
Don Huff, Huff Strategy & Communications
Monica E. Kuhn, Architect
Peter C. Lowitt, Devens Enterprises
Daniel K. Slone, McGuire Woods, LLP

Contents

Foreword: Arthur Erickson and Cornelia Hahn Oberlander 6

Introduction: The Rise of the Green Roof Industry in North America: Toward a Living Architecture 8

The Evolution of Green Roofs: From Their Origins to Excellence 16

Residential Projects and Award Winners
    Solaire Building – Balmori Associates, Inc. 30
    Island House – Shim Sutcliffe Architects ...34
    North Beach Place – PGAdesign^{inc.} Landscape Architects 38
    Yorktowne Square Condominiums – Building Logics, Inc. 42
    Eastern Village Co-housing Condominiums - EDG Architects 46
    Seapointe Village Deck Restoration – Jeffrey L. Bruce & Company 50
    10th@ Hoyt Apartments – Koch Landscape Architecture 54
    Lot 8 Santa Lucia Preserve – Rana Creek 58
    The Louisa – Walker Macy 62

Institutional Projects and Award Winners
    Ducks Unlimited National Headquarters & Oak Hammock Marsh Interpretive Centre – Number Ten Architectural Group 68
    Peggy Notebaert Nature Museum – Conservation Design Forum 72
    The Church of Jesus Christ of Latter-Day Saints Convention Center – Olin Partnership 78
    Oaklyn Library, Evansville Vanderburgh Public Library – Roofscapes, Inc. 82
    Life Expression Wellness Center – Roofscapes, Inc. 86
    Schwab Rehabilitation Hospital – American Hydrotech, Inc. 90
    Evergreen State College Seminar II Building – Garland Company 94
    Ballard Branch of the Seattle Public Library – American Hydrotech, Inc. 98
    The Green Institute (Phillips Eco-Enterprise Center) – The Kestrel Design Group, Inc. 102
    Mashantucket Pequot Museum and Research Center – Mashantucket Pequot Tribal Nation 106
    Nashville Public Square – Hawkins Partnership 110
    Sanitation District N°1 – Sanitation District N°1 of Northern Kentucky 114

Industrial/Commercial Projects and Award Winners
    901 Cherry Street – William McDonough + Partners 118
    Montgomery Park – Katrin Scholz Barth Consulting 122
    Garden Room – Buettner and Associates 126
    Burnham Park – Jeffrey L. Bruce & Company 130
    Ford Rouge Dearborn Truck Plant – William McDonough + Partners 134
    Millennium Park – Terry Guen Design Associates 138
    Heinz 57 Center – Roofscapes, Inc. 142
    601 Congress Street, Seaport District – Sasaki Associates, Inc. 146
    ABN AMRO Plaza – Barrett Company 150
    Calamos Investments – Intrinsic Landscaping, Inc. 154

Civic Awards: Championing the Cause
    Mayor Richard Daley, The City of Chicago 158
    Tom Liptan, The City of Portland 159
    Karen Moyer, The City of Waterloo 160
    Deputy Mayor Joe Pantalone, The City of Toronto 161
    Council Member Lisa Goodman, The City of Minneapolis 163

Research Award: Developing Performance Knowledge
    Dr. David Beattie, Founder of the Penn State Center for Green Roof Research 164

Bibliography 165

Appendix A: About Green Roofs for Healthy Cities 166

Appendix B: Plant Lists 168

Appendix C: Standards 172

Appendix D: Green Roofs and Leadership in Energy Efficient Design (LEED®) 173

Appendix E: Selected References and Resources 176

Foreword

> *"Green roofs are integral to the raison d'etre, concept, and vision of the architecture and landscape architecture as an inseparable whole. A green roof is not an 'addition' to the work, but integral to it."*

For over twenty-five years, we have been working together to bring out the very best in architecture and landscape architecture with our many engineering colleagues. Our landmark project, Robson Square, completed 1983, is an early testament to the power of collaborative green roof design.

Robson Square is a three-block project incorporating a courthouse, offices, an ice rink, and art gallery, in the busy centre of the City of Vancouver, British Columbia. It features waterfalls, pools, plazas, trees, and grassy mounds of rhododendrons. A linear urban park, Robson Square incorporates one of the first major uses of intensive green roofs in North America.

Primarily from Germany and spearheaded by Green Roofs for Healthy Cities in North America, a whole new industry has recently emerged to provide building blocks for a new generation of green roof designers – waterproof membranes, drainage layers, specialized irrigation systems, modular systems, and specialized growing media and plants.

Back in 1974, when we designed Robson Square, we had very little industry innovation or expertise to support it. The project demanded new research in lightweight growing media, plant material suitable for urban areas, drainage, voiding, and waterproofing. It also required innovations in technology and engineering, while incorporating ecologically sensitive and sustainable design strategies long before such matters became the civic concerns they are today.

Our philosophy, as innovators, thinkers, and pioneers, may be summed up as: "Taking a structural approach to landscape and a landscape approach to architecture." This requires that we respond to site conditions and light, which are integral to the landscape, and conceived of at the same time as the structure. This approach allows for unlimited opportunities for integrated design, as so many of the award-winning projects within this book beautifully demonstrate.

We have championed green roofs as a concept from the beginning of our long collaborative relationships, now encompassing over three-dozen projects. Green roofs are integral to the *raison d'etre*, concept, and vision of the architecture and landscape architecture as an inseparable whole. A green roof is not an "addition" to the work, but integral to it.

Robson Square is an open green civic space, which brings pleasure to its many visitors. It raises the oxygen levels and cools the city, reduces dust and germs; the waterfalls subdue the noise of the city traffic, and finally this complex gives the citizens of Vancouver a joyful meeting place. Today, the provincial government building at Robson Square is undergoing a renovation, to replace aging membranes and drainage, retaining the original trees and shrubs. Its restorative transformation is not unlike the major improvements in the green roof industry over the past five years.

Today with climate change challenges, we know that green roofs offer solutions to many of our urban health and density issues. This book illustrates many recent award-winning examples of creative and exciting green roofs leading the way to healthier cities and a brighter future for all.

We hope that our collective contribution and the projects herein, provide you with inspiration and we welcome continued exploration of the role of green roofs in our cities and lives.

Arthur Charles Erickson, CCBArch,
FRAIC, FAIA (hon), FRAIS (hon)
Cornelia Hahn Oberlander,
C.M., FCSLA, FASLA
2007

Foreword 7

Bird's eye view of intensive green roof, Robson Square, Vancouver British Columbia. *Courtesy of: Cornelia Hahn Oberlander*

Quiet area on Robson Square with Azaleas, Japanese Maples, Pines, and benches. *Courtesy of: Cornelia Hahn Oberlander*

Introduction:
The Rise of the Green Roof Industry in North America: Toward a Living Architecture

In early 1997, I sat beside a man named Dr. Brad Bass, a scientist with Environment Canada. We were both attending a conference organized to identify best practices in financing local government environmental initiatives – and it was lunchtime. Little did I know that our discussion would change the course of my life.

Between courses, Dr. Bass told me about something he called "green roofs." These green roofs, he said, were well established in Germany, yet virtually unknown in North America. He told me they could help manage stormwater, filter air pollutants, and save energy. If we built enough of them, he said, we might even improve the health of entire communities.

I was skeptical.

I had just spent a decade consulting on policy-development to support the diffusion of new environmental technologies. I was also working on identifying barriers to more sustainable community development, such as the need for more integrated land use and transportation planning. Could this green roof technology really do all that Dr. Bass suggested?

Despite my initial doubts, I felt compelled to learn more – and the more I learned about green roofs, the more I became excited with their tremendous potential. For here was an environmental technology that could not only contribute to community sustainability but one also holding the promise of being able to positively transform our lives on a massive scale – to lighten our ecological footprint.

Within a month of that first meeting, Dr. Bass and I combined efforts to apply for a grant to study the quantifiable benefits of green roofs and green walls – and most importantly, the barriers to their widespread adoption. The following year, with $25,000 of grant money in hand, we partnered with Monica Kuhn, a Toronto-based, German-speaking architect and local green roof advocate, and started our research.

One of our first tasks was to catalogue and understand the extensive body of literature and research already published in Germany (also exclusively written in German). There were dozens of meetings with North American pioneers, such as leading horticulturalist and researcher, Marie Anne Boivin of Soprema Inc., a France-based roofing company that was among the first to invest in green roof research in North America. We talked with many innovative policymakers, including Tom Liptan in Portland, Oregon, who began his own green roof stormwater research on his garage around 1997. We also organized a workshop in Toronto, Canada attended by more than 100 designers and policymakers. A handful of industry representatives turned out, helping us identify barriers to progress and dozens of suggestions for action. All of this research culminated in an exhaustive report entitled: *Greenbacks from Green Roofs*, published in 1999 by the Canada Mortgage and Housing Corporation, a federal government agency overseeing housing issues.

Greenbacks from Green Roofs was essentially a blueprint for the development of the green roof industry. We examined a broad range of evidence as to the public and private benefits of green roofs, while identifying barriers to their widespread implementation, including:

Lack of design and installation knowledge in the marketplace;
Lack of demonstration projects and experience with the technology;
Lack of North American research on the performance benefits;
Absence of public policy support; and
Additional capital costs of green roofs compared to traditional roofing.

> *"What if humans designed products and systems that celebrate the abundance of human creativity, culture, and productivity? That are also so intelligent and safe, our species leaves an ecological footprint to delight in, not lament?"*
>
> — William McDonough and Michael Braungart, *Cradle to Cradle*

Fortunately, we knew such barriers are typical for many new technologies. At the time, we had the benefit of considerable research, product development, and green roof expertise from Germany and other European countries to draw upon, but virtually zero information from North America. There was even less substantive research or evidence available from German sources on green walls. In fact, there wasn't a single company marketing green wall technology in North America.

I had become tired with writing reports about new policies and programs that often ended up languishing on the desks of bureaucrats. Clearly there was a need for the early pioneers in research, design, public policy, manufacturing, and installation to work together, to combine forces and resources to begin to tackle these challenges and proactively take action, with or without government support.

And so in 1999, I prepared a business plan to establish a coalition to focus on developing research and demonstration projects for a variety of green roofs in Toronto – and pitched it to ten interested companies. Seven agreed – and Green Roofs for Healthy Cities was born.

Our mission at Green Roofs for Healthy Cities is to increase the awareness of the economic, social, and environmental benefits of green roofs and green walls, and other forms of living architecture through education, advocacy, professional development, and celebrations of excellence.

Our initial goal was to aim high, to build a green roof atop Toronto's City Hall. Our vision was that it would need to be publicly accessible as a showcase of different technologies. To convince skittish City Councillors the project was worthwhile, we constructed a temporary 20 by 20 foot extensive green roof over some of the many concrete pavers that cover more than 70,000 square feet of roof terrace around City Hall's two towers. This temporary green roof sent a strong and graphic message as to how this huge area of downtown Toronto was largely being wasted. The temporary green roof and support from some key Council members such as then Deputy Mayor Case Ootes, helped to clinch the deal. On November 2, 2000, we formally launched the Green Roof Research and Demonstration Project, with the completion of two green roofs on Toronto's City Hall and the Eastview Neighbourhood Community Centre, a city-owned building that helped us gather data on stormwater retention and energy savings.

Founding Members of Green Roofs for Healthy Cities, an Industry Coalition Formed in 1999:

Doug Flynn, Flynn Roofing
Colin Donaldson, Soprema
Bill Bean and Brian Lambert, Garland Company
Bill Stensson, Sheridan Nurseries
Kaaren Pearce, DeBoer Landscaping and Maintenance
Al Duwyn, IRC Building Sciences Group
Frank Baxter, Semple Gooder Roofing

Butterfly garden, one of eight plots on the City of Toronto Green Roof Research and Demonstration Project, 2001. *Courtesy of: Steven W. Peck, Green Roofs for Healthy Cities*

The City Hall green roof demonstration project was comprised of eight publicly accessible rectangular sections with two green roof systems from Garland and Soprema. The intent was to showcase the variety of green roof applications so we included: two extensive, lightweight sedum-based systems; two semi-intensive systems, one of which was designed to replicate the black oak prairie ecosystem type, a kitchen garden, a butterfly garden, and a permaculture garden with corn, beans, and squash. Monitors for energy flows through the roof assembly were installed by the National Research Council's Institute for Research in Construction at both sites and we began to gather valuable data. (Liu and Baskaran 2003)

Designing, constructing, and monitoring these first two green roofs, was tough – we were, after all, breaking new ground. The project involved multiple funding sources – each with different requirements and different researchers – as well as public and private stakeholders, all of whom were not completely sold on the idea. In completing these projects we advanced our technical knowledge of the energy and stormwater benefits of green roofs, and educated policy makers and the public through regular tours and media coverage.

During this time, Green Roofs for Healthy Cities also began to study how green roofs could reduce the urban heat island effect, an increasingly serious problem faced by cities worldwide. Rooftops and paved surfaces have replaced vegetation in our cities and convert the sun's rays into heat during the day, which also builds up in many materials over time, resulting in significantly higher urban and suburban temperatures 2-10°F (1-6°C) relative to the surrounding countryside. Higher temperatures in cities mean more air pollution, hundreds of millions of dollars in additional energy consumption, a degraded quality of life, and even, during extreme temperature events, loss of life.

Green roofs provide public and private benefits ranging from the product itself (e.g., use of recycled and recyclable materials) to the region and global level (e.g., preservation of regional biodiversity and reduction in the quantities of green house gases). *Courtesy of: Green Roofs for Healthy Cities*

Here again, my friend and colleague Dr. Bass undertook the challenge by conducting the first green roof urban heat island study in the world. Dr. Bass' research clearly demonstrated how green roofs implemented widely, would help cool cities, save energy, and reduce air pollution, particularly the formation of ground level ozone, which increases with rising temperatures. This project was very important, since it was the first, citywide green roof benefit study ever conducted in North America. It opened up an entirely new area of ongoing research on green roof benefits, including stormwater management, energy, and air quality improvements at a broader community-wide scale.

With our growing profile, it was only natural that Green Roofs for Healthy Cities began to receive more inquiries about our work, particularly from manufacturers, researchers, and designers across North America and even Australia and the United Kingdom. Harnessing this growing interest, we opened up the coalition to additional members in 2001, and began to organize workshops in cities like Vancouver, Ottawa, and San Francisco. With these events, our little coalition began to mushroom with new and interesting webs of connections.

For instance, at the 2001 Ottawa workshop, hosted by the National Research Council's Institute for Research in Construction, we learned about green roof research that had been started in Ottawa under the leadership of Dr. Karen Liu. The Ottawa workshop was also attended by Dr. David Beattie, who would go on to develop the first green roof research center in the United States at Penn State University. In Ottawa I also met Ed Snodgrass, who would develop the first, fully dedicated green roof nursery specializing in sedum production in Maryland. Ed Snodgrass, one of the leaders in the field of green roof horticulture, recently published a definitive guide on extensive plants called *Green Roof Plants: A Resource and Planting Guide*.

The Vancouver workshop in 2001 laid the groundwork for the establishment of a green roof research facility at the British Columbia Institute of Technology spearheaded by Maureen Connelly, whose ongoing research in the field of noise reduction continues to advance our technical knowledge of this important green roof benefit.

With this exciting momentum, we determined it was time to mount a full-scale green roof conference. I approached Chicago's Mayor Richard Daley and his staff – public policy leaders in the promotion of green roofs – for their support in helping us co-host the first North America Greening Rooftops for Sustainable Communities conference. They agreed to help and we worked closely with Kimberly Worthington and Kevin Lebarge of the City of Chicago's Environmental Department to arrange facilities, develop walking tours, share technical information from their research, and provide a speaking opportunity for Mayor Daley. Over 400 people from around the world gathered to network, share information, and do green roof business. Our first plenary meeting was in the famous, ornately decorated Gold Room in the Congress Plaza Hotel on Michigan Ave. A fitting setting for what many rightly perceived as a significant business opportunity.

Mayor Daley won our first *Green Roof Civic Award of Excellence*. Upon accepting the award, I'll never forget how Mayor Daley appeared to be surprised over the fuss, especially from the local media, as he had been fighting a seemingly uphill battle in his own efforts to develop a green roof demonstration project at Chicago's City Hall. In fact, today we are proud to note that Chicago continues to support the building of more green roofs than any other city in North America, with more than 2.5 million square feet completed.

Needless to say, staging this first conference was a huge milestone. However, as is always the case with any steep growth curve, the event also presented our industry with a new challenge picked up in our *Greenbacks* report. Widespread discussions revealed the need to develop a professional accreditation program, with training courses and a certification for an *Accredited Green Roof Professional*. Given the inherent multi-disciplinary nature of green roofing, and the many opportunities for failures during the design and construction process, we collectively realized that the development of an *Accredited Green Roof Professional* designation would help protect the industry from the inevitable failures and provide a way for experts to shine with professionalism in the marketplace. The training program would also provide us with the opportunity to share the growing performance data provided by blossoming green roof research with the designers that need it.

Over the past five years, a significant body of research on green roof performance has been generated. Under the leadership of Michael Gibbons, new standards and guidelines for green roofs were established by the American Society for Testing and Materials (ASTM). With more field experience gained, Green Roofs for Healthy Cities has been able to harness the expertise of a broad range of professionals drawn from different disciplines, including roofing consultants, horticulturalists, architects, landscape architects, scientists, engineers, and manufacturers. They have been the cornerstone enabling us to develop consensus on best practices in a number of key areas such as project management, waterproofing, structural engineering, drainage, growing media, and, naturally, plants. This information has been assembled into four full-day courses that cover the broad range of design and implementation issues that are critical to success. *(See Appendix A)* We have also developed a half-day course that is focused on green roof design for ecological restoration

purposes. In 2007, we invited seventeen subject matter experts to come together in Toronto, and virtually locked them in a room until they agreed on the precise knowledge and skills required to become an *Accredited Green Roof Professional*.

Thanks to the tremendous effort of these and other dedicated professionals working hundreds of volunteer hours to establish a core body of knowledge in this rapidly evolving, multi-disciplinary field, we're close to realizing the goal we set in 2003. Accreditation will become a reality in the spring of 2009.

As mentioned, our mandate from the beginning included a mission to celebrate examples of design excellence in our field, so we established the *Green Roof Awards of Excellence*.

The objective of these awards is to honor the many outstanding contributions of organizations and individuals who are practicing integrative, multi-disciplinary green roof design and implementation. Other awards, such as the one presented to Mayor Richard Daley, recognize public sector individuals whose dedication and perseverance have made a significant contribution to the advancement of the industry through public policy.

Occupational Standard Development Participants:

Alex Johnston, Administrative Coordinator, Green Roofs for Healthy Cities
Charlie Miller, P.E., Principal, Roofscapes, Inc.
Chuck Friedrich, ASLA, Carolina Stalite, Co.
Dave Honza, Roofing Consultant, Honza Group, Inc.
Doug Fishburn, President, Fishburn Building Sciences Group
Jeffrey L. Bruce, FASLA, LEED, ASIC, President, Jeffrey L. Bruce & Company
Kelly Luckett, President, Green Roof Blocks
Kurt Horvath, President, Intrinsic Landscaping.
Michael Gibbons, FCI, President, Architectural Systems, Inc.
Monica Kuhn, OAA, Architect
Paul Kephart, Executive Director, Rana Creek
Robert Berghage, PhD, Associate Professor of Horticulture, Penn State University
Roger Schickedantz, AIA, Associate Partner, William McDonough + Partners
Stephen Teal, Architectural and Industrial Group, Flynn Canada
Steven Peck, Founder and President, Green Roofs for Healthy Cities
Steve Skinner, Garden Roof Project Manager, American Hydrotech, Inc.
Tim Barrett, President, Barrett Company
Virginia Russell, FASLA, LEED, University of Cincinnati

TOP 10 CITIES BY SQUARE FOOTAGE PLANTED IN 2006		
City	State/ Provence	Total Square Footage
1. Chicago	IL- Illinois	358,774
2. Washington, DC	DC- District of Columbia	301,751
3. Wildwood Crest	NJ- New Jersey	240,000
4. Dulles	VA- Virginia	230,000
5. Kansas City	MO- Missouri	178,008
6. Phoenix	AZ- Arizona	168,517
7. Milwaukee	WI- Wisconsin	79,513
8. New York City	NY- New York	67,896
9. Portland	OR- Oregon	64,442
10. Columbus	Oh- Ohio	58,025
Total Square Footage for 2006 (362 projects)		3,064,200

Annual survey of the square footage of green roof infrastructure implemented by a major city. *Courtesy of: Green Roofs for Healthy Cities*

A number of these jurisdictions, such as Chicago and Washington, have policies and programs in place which support green roof implementation, hence their leadership.

Since the inception of the *Green Roofs Awards of Excellence* in 2002, we have selected over thirty-five projects that illustrate an impressive array of green roof design innovation.

With this publication, we are proud to once again celebrate the excellent collaborative work of hundreds of green roof owners and design professionals who, by working in multidisciplinary teams, have stretched the boundaries of green roof and green building practice. But these boundaries are not impermeable. There is so much more that can be achieved through the integration of green roofs and green walls with other building systems to build healthier, stronger communities.

We need to better understand to what extent green roofs and walls can improve the performance of solar energy technologies; can diminish the need for air conditioning; and can utilize non-potable water sources. We need to know how these technologies may be configured to reflect regional climate differences, support our shared interests in preserving native flora and fauna, as well as contribute effectively to urban food production.

> *"Modern construction and development have also separated people from beneficial contacts with the natural environment, leaving the majority of urban residents to spend most of their waking hours in buildings lacking daylight, fresh air and exposure to nature. ...Creating less damaging, more positive relationships between nature and humanity however, will require dramatic changes in how we construct our buildings and landscapes."*
>
> — Stephen R. Kellert, *Building for Life: Designing and Understanding the Human-Nature Connection.*

Despite the many impressive innovations profiled in these pages and the explosion of green roof research over the last five years, we have only really just begun to explore the magnificent opportunities that green roofs provide.

In the future, we envision building projects of all shapes and sizes that not only provide shelter, places to live and work, but that also serve to heal and restore. We can create buildings that not only conserve resources, but also generate clean water and provide renewable energy to the grid. We can design buildings that process and manage waste; produce high quality food; and become incredible, physically and mentally healthy environments for all creatures.

If we are to thrive in this century and leave a better legacy for future generations, we need to adapt and brace ourselves for rapid resource depletion – declining supplies of oil and water, in particular – and an increasingly unstable political situation and an unpredictable climate. At this time, we need to embrace a new form of design practice like never before – a living architecture.

At the core of living architecture is a commitment to the integration of organic, living systems with the inorganic, lifelessness of modern building design. Living architecture requires a multi-disciplinary approach to building design and installation. Green roofs and green buildings are always at their best when they emerge from collaborative processes.

I believe that the green roof and wall building industry has the capacity to become a major force in support of sustainability, both locally and globally; and that it can meet that steep challenge within a generation. Green roofs and green walls must play a central role in that transformation.

Fortunately, we are well on our way.

Imagine if all of the rooftops and walls in your community supported plants? Imagine, for a moment, what the world would be like if all building projects completed from this day forward were restorative and healing? Would these buildings be good places to live? Work? Great places to raise a family or care for your grandchildren? Imagine what cumulative impacts living architecture would have on our communities. Just imagine it all – for this is the first step.

This book chronicles five years on the journey towards such a future – building-by-building, roof-by-roof. It contains outstanding examples of living architecture and shares the ideas and experiences from the professionals who brought them to life.

I hope you find inspiration from these remarkable people and projects and are able to take the second step – making it happen in your corner of the world.

Steven W. Peck
Founder and President,
Green Roofs for Healthy Cities

GREEN ROOFS
FOR HEALTHY CITIES

www.greenroofs.org

The Evolution of Green Roofs: From Their Origins to Excellence

> *"We look at architecture the wrong way, sideways, so what we see if just a thin sliver of the reality around us. To see architecture fully, you must stand it on its edge. When you do you always see dead land on display."*
> — Malcolm Wells, *Rediscovering America*

Reinventing Forgotten Spaces

Roofs are the forgotten "fifth façade" of our buildings. Mostly, they are ugly, barren places where we deposit refuse, stage billboards, and mount heating and cooling equipment and telecommunications towers. Indeed, our rooftops are truly the last urban frontier. As such, they represent an incredible opportunity to practice living architecture and, in so doing, reclaim the value of this underutilized real estate while healing our stressed planet.

The award-winning projects profiled in this book demonstrate how green roof technology provides a virtually limitless palette of design opportunities for innovators. Going forward, their imagination and ingenuity will transform hundreds, if not thousands of miles of wasted and forgotten roof spaces.

Unparalled Public and Private Benefits

There isn't another set of comprehensive building technologies that match the many benefits of green roofs while simultaneously inspiring and fulfilling our deep longing for all things natural.

They come in all shapes and sizes, and provide numerous public and private benefits. Green roofs insulation qualities help to conserve and temper the extraordinary pull of energy required to run air conditioning and heating systems in commercial buildings and homes, thereby saving money. Green roofs reduce the intrusion of noise pollution around freeways and airports, like the 901 Cherry Street in San Bruno, California. They also increase the lifespan of waterproofing membranes by protecting these essential layers from the wear and tear of the elements in all climates, all the while reducing the enormous annual flow of re-roofing waste into our cities' already stressed landfills.

Each green roof project is unique, with performance closely correlated to a number of interlocking factors. Just how much energy is saved through the installation of a green roof depends on its size, the climate in which the building is situated, how much natural and man-made water is available, the depth and quality of growing medium and what kind of plants are used. Additional energy savings may be acquired by modeling the green roofs impact on the mechanical system to right-size the air conditioning units. Furthermore, on a green roof, the ambient air above the plants is cooled and can be drawn into the air intake systems of air conditioning systems to also reduce energy consumption. (Leonard and Leonard 2005)

Green roofs need not compete for space with solar panels. By integrating green roofs with solar photovoltaics, it is also possible to increase the efficiency of electricity production. (Kohler 2003) On-roof renewable energy solutions such as solar can further curb the generation of greenhouse gas emissions and local air pollutants associated with traditional forms of energy production such as coal and natural gas.

Green roofs provide soothing, calming places where a building's residents may reconnect with the land – a true oasis for horticultural therapy and relaxation. They are also being used as living research and education sites while providing positive marketing and community relation opportunities. On them, we can grow local, organic food for restaurant kitchens, helping to curb the greenhouse gas emissions generated when food is transported over long distances.

Green Roofs as Infrastructure

Green roofs directly and indirectly connect the private spaces of the built environment with the public spaces of the surrounding site and community. From a public benefit perspective, choosing to invest in the widespread establishment of green roof infrastructure in North America over the next twenty to twenty-five years, as so many other countries already are, will result in enormous and far-reaching positive impacts. More and more governments are producing green roof incentives in the market. In doing so, they are able to leverage private investment, and utilize wasted roof spaces to help achieve multiple human health and environmental objectives. Each year we survey our members in order to identify the square footage and the type of green roofs installed. This information allows us to track the implementation of green roof infrastructure in each city. The potential market for extensive green roofing on existing buildings in cities such as Toronto, is an estimated 500 million square feet. The long term potential for the growth of the industry as a new form of infrastructure is underscored by the fact that the total square footage planted in 2006 was just over 3 million square feet - for all of North America.

Research at Columbia University and by Environment Canada further demonstrates that when green roof infrastructure is implemented in sufficient numbers, green roofs can contribute significantly to cooling cities thereby reducing the urban heat island effect. (Rosenweig et al. 2005; Bass 2003)

How this is achieved is fairly simple. Green roofs turn hot rooftops into virtual 'outside' air conditioners through the process of evapotranspiration whereby the sun's energy is used to evaporate moisture in the growing media and help plants transpire as they photosynthesize. When cities become cooler everyone wins. Cooler cities are healthier, better places to live during the summers. Cooler cities mean less ground-level ozone and peak-load energy consumption related to air conditioning in addition to a reduced need for polluting power plants and new transmission lines.

When combined with other measures, such as urban forestry and reflective pavement, green roofs can rapidly accelerate the cooling of entire cities. A five-year study by Environment Canada correlated summer temperatures and energy consumption patterns in Ontario, Canada. Researchers found that each 1.8°F (1°C) reduction in summer temperatures would shave approximately 4 per cent off the peak demand for electricity. Cooling the city would result in tens of millions of dollars in energy cost savings, not to mention less air pollution such as nitrous oxide and fine particulate matter in the air as a result of the increase in leafy plants. (Doshi et al. 2005)

In 2006, a report prepared for the New York Energy Research and Development Authority by the Columbia Center for Climate Systems Research explored opportunities to reduce New York City's urban heat island. The study utilized a regional climate model in combination with observed meterological satellite and GIS data to determine the impact of urban forestry, green roof, and light coloured surfaces on the urban heat island. During the summer months the daily minimum temperature in New York City is 7°F (4°C) warmer than surrounding rural and suburban areas. Nine mitigation scenarios were evaluated and six citywide case study areas. The results indicate that vegetation,

rather than albedo or other features of the urban physical geography, such as road density, was crucial in determining the urban heat potential in New York City. The report concludes that a combined strategy, which maximizes the amount of vegetation in New York by planting trees along streets and in open spaces and building green roofs offers more potential cooling than any one individual strategy. (Rosenzweig et al. 2005)

Greening more of our rooftops, especially in densely populated urban areas, will facilitate a multitude of infrastructure benefits that solve existing problems we already pour billions of public dollars into each year, such as the production and distribution of energy, as well as the storage and treatment of stormwater.

In fact, the capacity for green roofs to retain stormwater is already helping many clients meet their jurisdiction's regulatory requirements for onsite storage. Like other aspects of green roofs, the capacity for stormwater retention is always dependent on a number of interconnected design factors including the depth of growing media, roof slope, plant type and coverage, and whether or not the drainage layer is designed to store water in the first place.

Green roof plants and growing media can also help to improve the quality of stormwater runoff, reducing heavy metals in the atmosphere, itself the negative side-effect of traditional energy production and other industrial and transportation sources.

> **Various Policies used to Support Green Roof Implementation:**
>
> A number of cities are developing policies in support of green roof infrastructure, including:
> **Regulatory Measures:**
> Green building standards
> Fast track permitting
> Green space allocation/green area factor
> Green procurement by government facilities
> **Financial Incentives:**
> *Direct*
> Density bonusing
> Direct investment/grants
> Tax credits
> *Indirect*
> Low interest loans
> Energy efficiency incentives
> Storm water fee rebates

Engineering studies in Portland, Oregon, Washington, D.C., and Toronto, Ontario demonstrate how widespread green roof investment has the potential to save millions of dollars in public investment through the necessity for fewer and smaller public stormwater collection tanks, as well as a reduced demand for expensive erosion control infrastructure in urban streams and rivers.

A study funded by the Environmental Protection Authority, Office of Water, was conducted to determine the storm water benenfits of trees and green roofs in Washington, D.C. The study was prepared by Casey Trees, a non-profit organization dedicated to the promotion of urban forestry in Washington, D.C., and Limno-tech, an engineering firm. Two scenarios were used to determine incremental increases in tree and green roof coverage and studies with the D.C. Water and Sewer Authority hydrologic and hydrolic model. The first "high-end" scenarios considered putting trees and

green roofs wherever it was physically possible. Two scenarios were run for an average year (1990) wet weather continuous simulation, a six hour one inch design storm. Findings show reductions in storm water volume up to 10 per cent across the city and up to 54 per cent decrease in individual sewer sheds. As a result, planning is underway to use trees and green roofs as an integral component to satisfy D.C.'s storm water management strategy. (Casey Trees 2005)

A 2005 multi-disciplinary study led by Hitesh Doshi of Ryerson University in Toronto has helped to quantify the economic benefits of widespread green roof implementation. Eight per cent extensive green roof coverage of existing roofs in Toronto resulted in more than $300 million in capital infrastructure savings due to increased energy efficiency (urban heat island and direct savings) and improved stormwater management as well as just under $40 million in operational savings annually. (Doshi et al. 2005) Green Roofs for Healthy Cities estimates that reaching these levels of green roof coverage will likely take twenty-five years or more, and require public investment in the neighbourhood of $30 to $40 million per year in order to leverage the required private investment.

Green roof projects in this book, such as Yorktowne Square, the Ford Rouge Dearborn Truck Plant, Sanitation District No.1, and Life Expression Wellness Center are just a few outstanding examples of the possibilities for integrated water management designs. These projects not only improved stormwater management processes, but in several cases, have facilitated the harvesting of stormwater and graywater for green roof irrigation purposes.

Continued research demonstrates that reductions in stormwater runoff, thanks to a green roof, can also lead to a decrease in the frequency of combined sewage overflows in older cities with aging infrastructures, helping to keep beaches clean and open during the hot summer months by preventing diluted sewage from reaching their shores.

Numerous studies demonstrate how harvesting stormwater effectively through green roof infrastructure may also promote cost savings related to the need for less potable water creation for irrigational purposes.

> "Why not enlist the roofs of the world's cities in the campaign to save species? Who knows how many we could help if we varied the design of the gradens? In addition to growing them in the world's many climates, we could vary their soil and their exoposure to sun. We could even burn some gardens periodically to protect the species adjusted to such distances."
>
> — Michael L. Rosenzweig, *Win-Win Ecology: How the Earth's Species Can Survive in the Midst of Human Enterprise.*

Ecological Restoration: The Sky's the Limit

The rapid and unexplained loss of pollinators captured the media's attention in 2007, and raised alarms about the importance of bees and other species in the continuation of plant and human life. As our urban boundaries relentlessly expand, habitat and farmland continue to disappear at an alarming rate. Ecological restoration is relatively new and represents an important new research field and industry. Researchers and practitioners in this field have yet, for the most part, to fully acknowledge the immense potential of green roofs to contribute to the survival of endangered and threatened species – but the circles of awareness expand every day.

Consider the pioneering work of Stephan Brenneisen in Switzerland and his relentless efforts to implement green roofs that promote biodiversity as a primary design objective. In Switzerland, green roofs are being designed with a view to replacing the lost habitat of specific species. In London, United Kingdom, Dusty Gedge's passion to protect the Black Redstart bird has led to the development of over one million square feet of green roofs that support their declining and eroded habitat, some of which adorn the Canary Wharf complex. In the United States, living architecture icon Paul Kephart of Rana Creek, has led the charge with his green roof designs in support of the rare Bay Checkerspot Butterfly.

The award-winning Oak Hammock Marsh and Feldman Residence projects in this book further demonstrate what can be accomplished in two very different North American ecosystems. Beautifully designed, these green roofs provide habitat for a variety of Midwestern Prairie and Central Californian species, helping the buildings blend into the landscape. There is an ever-expanding syllabus of scientific evidence and real-life projects that prove, once again, how green roofs, green walls, and other forms of living architecture can provide protected habitat for rare and endangered plants and various species of invertebrates.

Green roof designed to enhance biodiversity, Basel, Switzerland.
Courtesy of: Steven W. Peck, Green Roofs for Healthy Cities

With so many new research projects being undertaken every day, in virtually every corner of the world, we expect the body of knowledge pertaining to the tangible benefits of green roofs' capacity to increase biodiversity to continue to expand exponentially – indeed, it is the very *raison d'etre* for the online database of research project summaries and policies called *The Green Roof Tree of Knowledge*.

The greening of rooftops can also address some of the central goals of the so-called "low-impact development" and "smart growth" movements, both concerned with learning to design more efficient, compact, and livable communities. Anecdotal evidence from developers suggests that green roof projects help reduce community opposition to infill development projects by turning inhospitable black roofs into useable, enjoyable resources for the community to visit or simply see. Several governments, most notably Chicago and Washington, provide developers with a 'fast track' permitting option if developers include specified green building features in their new developments. Others, like Portland offer floor-area density bonuses in exchange for green roof implementation.

Sustainable communities require the redevelopment of cities in a manner that increases population density while also maintaining a high-quality of urban life. Within the pages of this book you will find examples of children's playgrounds at North Beach Place in San Francisco, horticultural therapy gardens in the Schwab Rehabilitation Centre in Chicago, and fabulous places to relax and simply enjoy, like The Solaire in Battery Park City, New York.

Achieving the goals of more livable, compact urban spaces through more efficient and logical use of rooftop greening principles is perhaps best captured through the realization of Chicago Mayor Richard Daley's twenty-four-acre Millennium Park. Unveiled in the summer of 2004, this astounding project is among the largest and most complex rooftop parks ever completed. Millennium Park also represents a breakthrough in city building and planning through the reclamation of an ugly and under-utilized transportation corridor next to Lake Michigan, turning it into a world-class park and natural oasis for enjoyment by all.

Though award-winning green roof design provides multiple private and public benefits, not all benefits can be realized in every project, because there are always tradeoffs. The many design objectives realized by this collection of projects are listed in the table on this page.

With the advance of every new green roof, and more recently, every new green wall technology, every single roof and wall has the potential to contribute multiple benefits to building owners and the surrounding community.

Green Roof Design Objectives and Benefits

Aesthetic improvement
- Integration of art objects
- Four season aesthetic

Waste diversion
Storm water management quantity and quality
Reduction of the urban heat island effect
New amenity spaces
- Recreation spaces
- Horticultural therapy
- Food production
- Commercial uses

Local job creation
Energy efficiency
Increased membrane durability
Fire retardation
Noise reduction
Marketability
Ecological restoration
Improved health and well-being
- Improved air quality

Visual integration with the surrounding environment
Integration with other building components
- Photovoltaic panels
- Heating ventilation air conditioning units
- Other stormwater retention features
- Green walls
- Elimination of internal drains
- Rainwater harvesting for irrigation
- Greywater harvesting

Research applications
Educational opportunities
Community involvement
Low maintenance
Ecologically sensitive maintenance

Sedum are the most commonly used vegetation on extensive green roofs due to their resistance to the elements and variety of colors. *Courtesy of: Steven W. Peck, Green Roofs for Healthy Cities*

The Evolution: From Ancient Roof Gardens to High-Tech Extensive Green Roofing

Green roofs are not a new phenomenon. They have been a standard construction practice in many countries for hundreds, if not thousands, of years, mainly due to the excellent insulating qualities of the combined plant and soil layers (sod). In the cold climates of Iceland and Scandinavia, for instance, sod roofs help retain a building's heat, while in hot countries such as Tanzania, they keep buildings cool.

The first known historical reference to green roofs is on stone temples (*ziggurats*), built in ancient Mesopotamia from the fourth millennium until about 600 BC. The Hanging Gardens of Babylon is one of the most famous examples of these luxurious roof garden temples. Later, in the Viking Age (800-1000 AD) sod roofs became prevalent throughout Northern Europe. Turf, and occasionally seaweed, was used to line the walls and roofs of homes for protection from harsh winds, extreme cold, and rain.

Renewed European interest in this concept beginning in the 1960s in Germany, Switzerland, Austria, and Norway can be attributed to a growing concern over the degraded quality of life in the urban environment and the rapid decline of green space in intensely developed cities and towns.

The early German green-roof movement has roots in a deep discontent with the aesthetic of the post-World War II built environment. Individuals such as Hans-Joachim Liesecke, called for the greening of unused gravel areas to improve microclimates and building aesthetics. In the mid-1970s, research at the University of Berlin by botanist Reinhard Bornkamm, helped pioneer the development of what we now know as the extensive green roof system. Such systems are comprised of less than six inches of aggregate-based growing media and planted with extremely tough, drought-tolerant plant species, primarily *sedums*.

Extensive roofing systems radically differ from older "intensive" roof gardens in that they are designed for minimal maintenance and are extremely lightweight. Extensive systems were heavily studied by various German academic institutions starting in the 1980s and were found to have positive attributes in terms of stormwater management, plant survivability, fire retardation, and energy conservation. Important long-term studies of green roofs, and more recently, green walls, are still being carried out by a second generation of researchers such as Dr. Manfred Kohler at Neubrandenberg University and Marco Schmidt at the University of Berlin.

In 1975, a nonprofit research society called *Forschungsgesellschaft Landschaftsentwicklung Landschaftsbau e V.* (or simply, FLL) was established to conduct research and set standards and guidelines for the entire German landscaping industry. A variety of experts in horticulture, engineering, landscape architecture, and building sciences began meeting in the late 1970s to discuss technical issues related to rooftop greening. In fact, they coined the terms "extensive" and "intensive" green roofs, and established detailed technical guidelines in areas such as root repellency, plant selection, and growing media composition and performance.

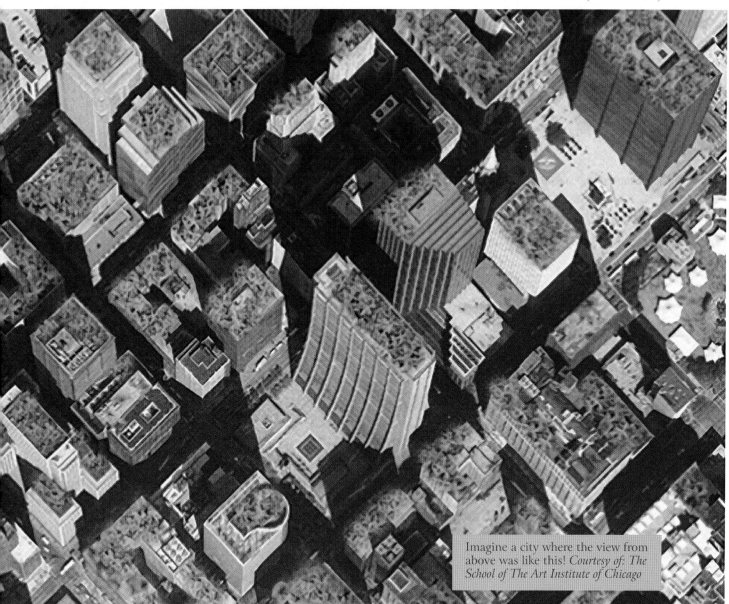

Imagine a city where the view from above was like this! *Courtesy of: The School of The Art Institute of Chicago*

The stage was now set.

The growing strength of the environmental movement in Germany in the 1980s, combined with pressing environmental issues such as stormwater management, groundwater pollution, and the urban heat island effect in many German cities became the backdrop for significant progress. The advocacy for roof greening by organizations such as the FLL, combined with the technical guidelines issued in 1982, and the ability of "extensive" lightweight green roofs to cover many existing buildings and address urban environmental challenges resulted in the implementation of supportive mandatory green-roof regulations and financial incentives for retrofit projects in major cities across Germany. These public and private partnerships help to forge what is now the modern green roof industry in Germany.

Today, free from the confines of significant structural loading capacity requirements, extensive green roofs have opened up the potential to retrofit billions of square feet on existing buildings. These new lightweight systems, combined with public policy support in over seventy-five local jurisdictions, have already resulted in an explosion of green roofing in Germany, Austria, Switzerland, and elsewhere in Europe.

It is estimated that the total German green roof area is approximately three billion-square feet, or approximately 15 per cent of all flat roof buildings with roughly five square miles being built on an annual basis. Given this, it will be no surprise that many of the products and services first introduced into the North American market have their origins firmly planted in this European experience.

Two modern advocates of green roof technology were the architects Le Corbusier and Frank Lloyd Wright. Although Le Corbusier encouraged rooftops as another location for urban green space while Wright used green roofs as a tool to integrate his buildings more closely with the landscape: neither was aware of the profound environmental and economic impact that this technology could have on the urban landscape.

Advocates with a more complete sense of the potential for the technology include Malcolm Wells and landscape architects and Cornelia Oberlander and Theodore Osmundson.

Malcolm Wells, often called the "father of modern earth-sheltered architecture," became an architect in 1953 and practiced for eleven years before deciding that contemporary architecture was destructive rather than creative; it destroyed the natural habitats that it was built within. Ever since then, Mr. Wells has been an advocate of buildings that can "heal the wounds caused by their construction," allowing plants and animals to return to blankets of living land within the built environment.

Cornelia Oberlander was the first woman to graduate with a degree in Landscape Architecture from Harvard in 1947. Her work has focused on improving the urban landscape and she is well recognized for her intensive green roof at Robson Square in downtown Vancouver as well as the extensive green roof which adorns the Vancouver Public Library. Oberlander has been a passionate supporter of green roofing, long advocating for increased public policy support for green roof technology in Canada and the United States. During a convocation speech where she accepted an honorary degree from Simon Fraser University in 2005 she stated:

> *"I dream of green cities with green buildings where rural and urban activities live in harmony…. 'Achieving a fit' between the built form and the land has been my dictum. This can only be done if all our design related professions collaborate and thereby demonstrate co-operatively their relevance in meeting the enormous developmental challenges facing our increasingly crowded urban regions."*

Theodore Osmundson, FASLA, is a well-respected landscape architect who began his practice in San Francisco in 1946. A past president of the International Federation of Landscape Architects and the American Society of Landscape Architects, he was the winner of the prestigious ASLA Medal in 1983.

Theodore was a keynote speaker at the first green roof event in the United States in San Francisco, California, in 2002 organized by the local Roofing Consultants Institute and Green Roofs for Healthy Cities. Recognized as a leading designer and advocate of intensive green roofs, Osmundson published an extraordinary book on his life's work, entitled *Roof Gardens: History, Design, and Construction* in 1997.

The history of green roofing is far from complete. New chapters are being written all the time and there remain many new areas to forge creative solutions to the challenge of integrating natural, living systems with the lifelessness of modern buildings.

Green Roof Basics

A green roof is defined as a green space created by adding layers of growing media and plants, root-repellent material, a drainage layer, filter cloth, and high-quality waterproofing on top of a roof that is located below, above, or at-grade.

Green roofs should not be confused with traditional terrace roof gardens on which planting is constructed in freestanding containers and planters located on an accessible roof terrace or deck; they are intimately connected to the waterproofing and drainage above the structural decking.

The essential layers of a basic green roof system from the top down include:

> The plants – components specially selected for particular applications;
> An engineered growing medium, which may not include soil;
> A landscape or filter cloth to contain the roots and the growing medium, while allowing for water penetration;
> A drainage layer, sometimes with built-in water reservoirs;
> A root repellent layer/component;
> The waterproofing membrane; and
> The roof structure with traditional insulation, either above or below.

Extensive green roofs are new to the market. Often, they are not accessible to the public and are characterized further by:

> Low weight;
> Low capital cost;
> Low plant diversity; and
> Minimal maintenance requirements.

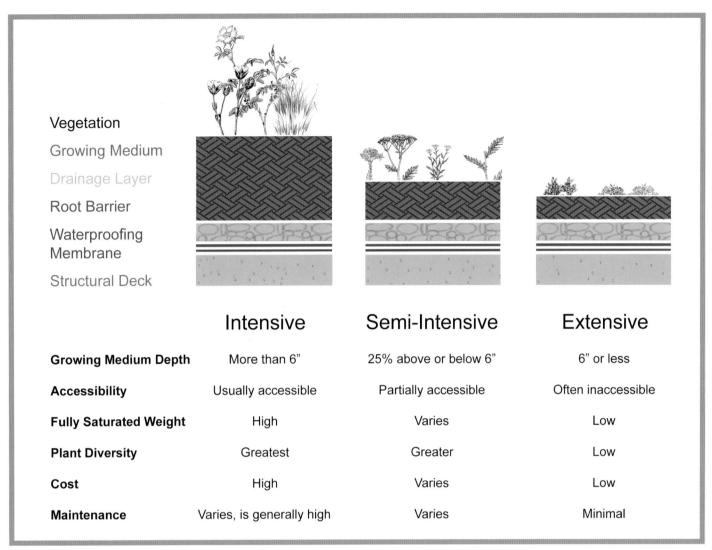

Green roof system overview and category attributes.
Courtesy of: Green Roofs for Healthy Cities

The growing medium is typically made up of a mineral-based mixture of sand; gravel; crushed brick; lightweight expanded slate, clay, or shale aggregate; volcanic rock; pumice stone; scoria; zeolite; diatomaceous earth; perlite; or rock wool. Organic matter may be comprised of: composted straw, saw dust, wood, grass, leaves, clippings, agricultural waste, worm castings, peat, peat moss or manure.

The growing medium on extensive green roofs varies in depth between 2-6" with a weight increase of between 12-35 pounds per square foot when fully saturated or at maximum density.

Due to the shallowness of the growing medium and the extreme desert-like microclimate on many roofs, plants must be low and hardy, typically alpine, dry-land or indigenous. The plants are often watered only until they are fully established and then, after the first year or two, maintenance will typically consisting of two visits a year for weeding of invasive species, and inspections of drainage and waterproofing membranes.

Intensive green roofs are roof gardens or parks on rooftops, almost always accessible to building occupants or the public, are characterized by:

 Deeper growing medium (more than 6 inches) and therefore greater weight;
 Higher capital costs;
 Increased plant diversity; and
 Higher maintenance requirements.

The growing medium is uniformly deeper than six inches with a saturated weight increase of between 50-300 lbs/sf and may include several blends, engineered to support different plants. Due to the increased growing medium depth, the plant

About The Green Roof Awards of Excellence

selection is more diverse and can include trees and shrubs.

Requirements for maintenance – especially watering – are more demanding and ongoing, and irrigation systems are usually specified for intensive systems. Consultation with an experienced structural engineer, horticulturist, and installer are particularly important for intensive green roofs with their deeper growing mediums, greater weight, and more diverse plant cover than extensive roofs.

Depending on such site-specific factors as location, structural capacity of the building, budget, client needs, and material and plant availability, each individual green roof will be unique – often a combination of both intensive and extensive systems.

Combinations of extensive and intensive – **'semi-intensive' green roofs** – combine growing media depths on the same roof, with each depth range accounting for at least 25 per cent of the roof's total area.

Semi-intensive green roofs are almost always accessible to building occupants or the general public, and their structural loading requirements vary.

Green roofs may be installed in separate layers, or in modules or mats that combine various elements together. Many manufactures offer complete systems, while others provide one or more of the essential components. Look for references to ASTM or FLL standards to substantiate product performance claims. Green roof design teams may elect to go with a complete system, which is often accompanied by guarantees, or to assemble and test the components of the system independently.

As the industry evolves, new products and services continue to come on the market, and this innovation is helping to drive down costs and improve efficiencies.

The award-winning projects featured in this book were chosen for their innovative green roof designs and ability to maximize multiple, potential benefits. Each year six outstanding green roof projects, three extensive and three intensive, are selected by a six-member, independent, multi-disciplinary judging committee that includes architects, landscape architects, engineers, and horticulturists. These generous individuals are recognized in the acknowledgements section.

Each submission is individually evaluated according to a broad range of weighted criteria including aesthetic, economic, functional, and ecological components. As you will see, each building type – residential, institutional, and commercial – is driven by a different design and performance goals. The top three submissions from each category were then discussed by judges and a winner selected. Our thanks are extended to all of the professionals who submitted their projects over the past six years.

In addition to awards for projects and their designers, there are also awards granted to individuals that have made an outstanding contribution to the development of the industry. The City of Chicago's Mayor Daley, The City of Portland's Tom Liptan, The City of Waterloo, Ontario's Karen Moyer, The City of Toronto's Deputy Mayor Joe Pantalone, and The City of Minneapolis' Council Member Lisa Goodman have all greatly contributed to the advancement and popularization of green roof technology through policy development.

Dr. David Beattie is the first and only recipient of the Research Award of Excellence for his establishment of the Penn State Center for Green Roof Research and his significant body of research on green roof performance benefits in North America.

Researchers, policymakers, green roof design professionals – we all have important roles to play in making green roofs and green walls a commonly accepted feature of living architectural practice in North America.

Residential Projects and Award Winners
Residential: Semi-intensive

Solaire Building

Location: Battery Park City, New York, New York
Project Type: New Construction
Green Roof Size: 5,000 square feet (intensive) and 4,800 square feet (extensive)
Completion Date: 2003
Year of Award: 2004

Client/Developer: Albanese Development Corporation
Design Architect: Cesar Pelli and Associates
Landscape Consultant: Balmori Associates, Inc. *(Winner)*
General Contractor: Turner Construction Company

Green Roof System Details
Structural Deck: Concrete.
Waterproofing: American Hydrotech, Inc. hot rubberized asphalt membrane in a Protected Membrane system.
Drainage: American Hydrotech, Inc. drainage, Gardendrain™, in different profiles for the extensive and intensive areas.
Irrigation: The intensive green roof has a permanent drip irrigation system while the extensive plot has no permanent irrigation. The irrigation system uses recycled water from a cistern in the basement.
Growing Medium: 6" - 12" depth of mixture of Northeast Solite mineral and organic material.
Vegetation: The intensive green roof was planted with a mixture of ground covers, grasses, perennials, shrubs, and bamboo trees. The extensive green roof was planted with a mixture of sedum plugs. (*See Appendix B: Plant Lists*)
Design Objectives: Storm water management quantity and quality; amenity spaces (recreation); energy efficiency; integration with other building components (heating ventilation air conditioning units, gray water harvesting, rain water harvesting for irrigation); local job creation.

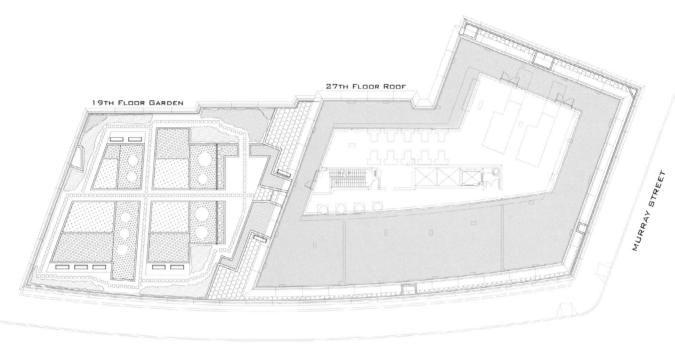

Plans for the 19th and 27th floor green roofs.
Courtesy of: Balmori Associates

The Solaire Building is the first 'green' residential high-rise in North America in response to an ambitious set of new guidelines for green architecture developed by the Battery Park City Authority. The building's two Balmori Associates, Inc. designed green roofs are an integral part of the sustainable, low-impact objectives of the entire building which received a LEED® Gold rating. *(See Appendix D: Green Roofs and Leadership in Energy Efficiency Design)*

The roofs are able to retain 70 per cent of the storm water that falls on their area, channeling all runoff into large cisterns for storage until needed for irrigation. Similarly, the green roofs are also irrigated with water from the building's gray water system. These two sources of water serve to eliminate the unnecessary use of potable water for irrigation. Likewise, air intake vents are carefully placed in the vegetated areas to take advantage of the cool ambient temperature.

The roofs not only serve a functional goal, but also warm the aesthetics of the building and enhance its environmental image. The often-accessed intensive 19th floor green roof overlooks the Hudson River and serves as an oasis from the noise of the streets below. It was planted with shrubs, perennials and bamboo trees, chosen for their aesthetic appeal, ability to resist drought and strong winds, and to adapt to shallow soil depths. The bamboo was strategically planted in the center of the garden to provide a year-round windscreen and shade the garden's pathways and benches.

Since receiving a Green Roof Award of Excellence, the green roof layout has been altered. The redesign, carried out by the developer, included the replacement of Bamboo with *Amelanchier* and the closing of some of the pathways. The second green roof is a sedum based extensive application and has remained largely unaltered.

At the time of implementation, this technology was not commonplace in New York. As a result, there were a limited number of contractors in the market who had experience or were familiar with the considerations of lightweight landscaping over materials like waterproofing membranes. This resulted in the need for increased supervision during the implementation process. These newly skilled landscaper contractors were then retained on warrantee for the establishment period.

Capacity development has proven to be quite useful since many other developments in the area are mimicking Solaire's design and building methods. This trend toward green building design has been brought on by Battery Park City's construction requirement and its high tenancy rate. Declared "One of the Ten Best Places to Live" in the United States by *Organic Style* magazine, at four months old the building had a 95 per cent occupancy rate with renters paying the equivalent of units in other luxury buildings (McLinden 2004). The green roofs and the building below have helped shift the nature of the archetypal city in which they are located: positively affecting the surrounding environment and well being of inhabitants.

"In their most modest and minimal form, green roofs offer an antidote to the impervious paved world in which we live and which we have very few chances to modify as part of an already built up city."

— Diana Balmori, Landscape and Urban Designer, Balmori Associates

Opposite: Bamboo bordered pathways on the 19th floor intensive green roof served to limit the impact of the wind on the end user. *Courtesy of: Balmori Associates*

June 2003 construction on the 19th floor intensive roof. *Courtesy of: Balmori Associates*

Residential: Extensive

Island House

Location: Thousand Islands, Ontario
Project Type: New Construction
Green Roof Size: 1,700 square feet
Completion Date: 2002-2003
Year of Award: 2004

Client/Developer: Carol and Kevin Reilly
Architect: Shim-Sutcliffe Architects *(Winner)*
Design Consultant: Mill & Ross Architects
Design Consultant: Donald Chong Studio

Green Roof System Details
　　Waterproofing: Soprema modified bitumen roof with a root repellent.
　　Drainage: Soprema drainage board.
　　Irrigation: No permanent irrigation system.
　　Growing Medium: 6" depth.
　　Vegetation: The upper green roof is a wildflower meadow with a mix of local indigenous flowers, which was later replaced with sedum species. The lower roof uses local sedum, including *Sedum album, S. floriferum 'Weihenstephaner Gold', S. kamtschaticum ellacombianum* and *S. Spectabile 'Brilliant'*.
　　Design Objectives: Visual integration with the surrounding environment; increased membrane durability; storm water management: quantity; local job creation.

Residential Projects and Award Winners

The wild flower meadow has since been replaced with sedum plantings to facilitate the maintenance process.
Courtesy of: Shim-Sutcliffe Architects

Unlike the majority of green roof projects, which are in urban areas, this single family home finds itself situated on a former dairy island on the St. Lawrence River between Ontario and New York State. Shim-Sutcliffe Architects approached the project with the intent of respecting the agrarian tradition of the region and maintaining the openness of the landscape while providing their clients with privacy and a splendid view of the river.

Before construction began, the five-acre site was hydroseeded with clover to reinvigorate the large meadow. This same mixture was then used on the upper rooftop in order to blend the structure into the landscape. The clover is harvested and transformed into hay bales by a local farmer, continuing the site's connection to its agricultural roots. However, though the green roof was aesthetically connected to the site, the vegetation in this elevated application proved to be problematic. Eventually the upper green roof was replanted with sedum species that were simpler to maintain. Similarly, on the lower roof plants were redistributed after the establishment period in order to create a stronger plant palette.

The careful attention paid to the vegetation after installation was the result of very careful planning. The architects and manufacturers worked closely together to develop the project. However, early in its development both realized that the distances from the nearest centers of green roof knowledge were too far to travel for maintenance personnel. As a result, a local was trained in basic maintenance procedures. This capacity development is in keeping with the projects goals of respecting the island and surrounding area's existing tradition of self-reliance.

The green roof is one of many landscaping elements that are used to link the structure to the landscape. On the St. Lawrence River side the house opens up to a large water garden with indigenous water lilies and bulrushes. This site redefines and balances the relationship between landscape, construction, and water.

"The green roof is part of larger vision for landscape; it is one part of a greater approach to the agrarian context of the island."

— Brigitte Shim, Architect,
Shim-Sutcliffe Architects

The roof creates visual unity between the house and its context. *Courtesy of: James Dow*

Sedum planted lower roof. *Courtesy of: Shim-Sutcliffe Architects*

Residential: Intensive

North Beach Place

Location: San Francisco, California
Project Type: New Construction
Green Roof Size: 1.9 acre
Completion Date: 2004
Year of Award: 2005

Client/Developer: BRIDGE Housing Corporation
Landscape Architect: PGAdesigninc Landscape Architects *(Winner)*
Building Architect: Barnhart Associates Architects
Associate Architect: FullCircle Design Group
Waterproofing Consultant: Simpson Gumpertz & Heger
Structural Engineer: FBA Structural Engineers

Green Roof System Details
Structural Deck: 13 1/2 inch project-wide set-down of the structural slab over post-tension structural slab.
Waterproofing: Grace Construction Products 60 mil modified bituminous sheet membrane.
Drainage: 3/8" Drainage Composite and Hydroduct 660.
Irrigation: 95 per cent overhead spray and 5 per cent drip emitters.
Growing Medium: 12-36" depth. Medium is divided in two layers, a top more organic layer supported by a layer of sandy aggregate.
Vegetation: 12 tree species, complimented by 60 other plant varieties. (*See Appendix B: Plant Lists*)
Design Objectives: New amenity spaces (recreation spaces).

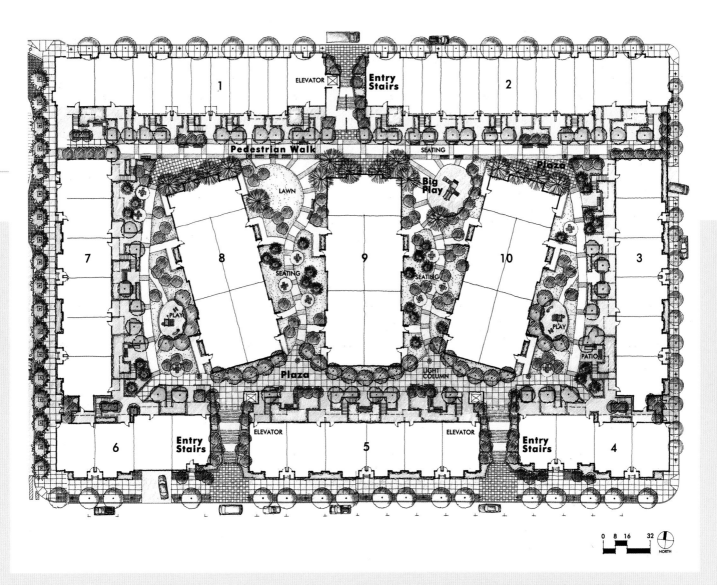

The landscape concept plan of the east block of North Beach Place shows small and large play areas for young and other children, as well as courtyard spaces for neighborhood use. *Courtesy of: PGAdesigninc.*

A park-like setting at unit entries. *Courtesy of: PGAdesigninc.*

Opposite: With the structural slab set down over a foot below the path surface, trees and other plantings have a generous reservoir of soil to grow in. *Courtesy of: PGAdesigninc.*

Land in San Francisco's upscale North Beach Neighborhood is a precious commodity. Through the North Beach Place project BRIDGE Housing, a developer who is dedicated to building quality affordable housing in the Bay Area sought to recapture valuable urban space for residents' use and enjoyment. The development replaces a two-block, 1950s public housing complex with a mixed-use development that includes 341 affordable family and senior apartments, 20,000 square feet of commercial space, childcare facilities, and underground parking.

Seven years in the making, this project shows that, with dedication, major funding issues can be overcome. North Beach Place was originally slated for nearly twice the federal funding it actually received; as a result, the developer was forced to rely on some creative financing strategies. Their dedication resulted in a partnership between five public lending bodies and one private lender.

A key component of the development is the second-story green roof. The green roof design, led by Cathy Garrett of PGAdesigninc. Landscape Architects, unifies all components of the project with a green oasis at the doorstep of each home. Residents have direct access to inviting outdoor spaces of different sizes and functions, including large and small play areas, seating, and places for community gathering. PGA's design took full advantage of the set down structural slab to minimize the height and visual impact of planter walls. The result is an intensive green roof with a park-like character that is comfortably human in scale.

In order to minimize through-structure penetrations and create a more flexible and dynamic ourdoor environment, Paul Barnhart, Barnhart Associates Architects, designed a project-wide, 13.5-inch set-down structural slab. This allowed the use of a reservoir of soil below the pavement elevation, resulting in generous volumes of growing medium in planting beds and the integration of a wide array of vegetation, including a dozen different tree species and sixty other plant varieties. In cases where the depth of the medium exceeded twenty-four inches, two growing media were used: one with organic content to support plant growth along with a base layer of fully inorganic mixture. Both layers were designed to be quick draining, but also to provide essential nutrients while retaining water in the growing medium.

The development was nearly fully occupied by July 2005, one measure of its success, but the pride that residents take in their renovated community may be a more appropriate index. The well-used park amenities of North Beach Place further demonstrate the extent to which it has become a focal point and a resource for the people it serves.

"North Beach Place is a model of affordable housing that not only fits the environment but also makes a substantial contribution to it."

— Cathy Garrett, Landscape Architect, PGAdesigninc. Landscape Architects

Residential: Extensive

Yorktowne Square Condominiums

Location: Falls Church, Virginia
Project Type: Retrofit
Green Roof Size: 4,700 square foot
Completion Date: 2003
Year of Award: 2005

Client/Developer: Yorktowne Square Unit Owners Association
Architect: Building Logics, Inc. *(Winner)*
Structural Engineer: Engineered Building Solutions, Inc.
Contractor: Paneko Construction, Inc.
Vegetation Design: Emory Knoll Farms

Green Roof System Details

Structural Deck: 3/4" tongue and groove plywood.
Waterproofing: EnviroTech green roof built up bitumen roof with root barrier.
Drainage: EnviroTech hydrogel packs.
Irrigation: No permanent irrigation.
Green Roof System: EnvironTech System.
Growing Medium: 2" depth of expanded slate and horticulture composite.
Vegetation: *Sedum album, Sedum reflexum, Sedum sexangulare.*
Design Objectives: Storm water management: quantity and quality; integration with other building components (rain water harvesting for irrigation); research applications; educational opportunities, Community involvement.

View of roof four years after completion.
Courtesy of: Building Logics, Inc.

The Yorktowne project serves as a model for residential and business communities demonstrating how green roofs and other storm water management designs can be implemented to improve declining water quality, decrease erosive storm water runoff, and conserve flora and wildlife resources in the Chesapeake Bay watershed. This green roof system was implemented as the first phase of an owner instigated storm water management strategy. The first phase of implementation was funded through a grant from the Environmental Protection Agency's Chesapeake Bay Program through the Virginia Department of Conservation and Recreation.

The fourteen unit condominium's original plywood roof deck was built in 1968. Its age and construction method did not lend itself well to a green roof. However, the client was dedicated to the endeavor and assumed the cost of the structural upgrade from clipped in place half-inch to three quarter inch tongue and groove plywood. As a result, the roof was able to support a dead load of fifteen pounds per square foot, allowing for an extensive green roof albeit a light one.

The owners employed a structural engineer to assess the carrying capacity of their roof, whose recommendations were implemented by local green roofer and architect Michael Perry of Building Logics. Perry, who had been recommended by the project's funder, manufactures a system that was compatible with the limited loading capacity. In addition to the typical green roof elements, this one included hydrogel packs. The packs, which are laminated to the waterproofing membrane, are composed of dry crystals that absorb water then turn it into a solid – much like a diaper. Plants seek out the moisture and grow into the pack through a geotextile layer. This effectively creates a monolithic plant covering. The hydrogel packs also serve to bolster the drainage system, which was not ideal, despite the installation of eight new cast iron drains with cast iron drain lines.

The rooftop water retention system is complemented by several at-grade cisterns that are being used to quantify the exact amount of storm water runoff from the green roof and an adjacent conventional roof. Initial monitoring by a George Mason University graduate student focused on storm water retention qualities, filtering characteristics, and temperature differentials. Reports confirm that 80 per cent of the annual rainfall has been retained on the roof. Since that time the site has been used by scientists at Florida State University and Michigan State University to research the water quality of green roof runoff.

Not only has this project fulfilled its owners' desire to reduce the impact of their building on the Chesapeake Bay watershed, it is being used to further understand and promote the ability of green roofs to do so on a wider scale.

Opposite: Sedum album two years after implementation.
Courtesy of: Building Logics, Inc.

View of roof four years after completion.
Courtesy of: Building Logics, Inc.

Residential: Extensive

Opposite: Green walls adorn the walls of the Eastern Village Cohousing Condominiums, providing continuous greenery for its residents.

Eastern Village Cohousing Condominiums

Location: Silver Spring, Maryland
Project Type: Retrofit
Green Roof Size: 8,000 square feet
Completion Date: 2005
Year of Award: 2006

Client/Developer: Eco Housing Corporation and Poretsky Building Group
Architect: EDG Architects
Environmental Building Consultant: Sustainable Design Consulting
Landscape Architect: Lila Fendrick Landscape Architecture & Garden Design

Green Roof System Details
  Structural Deck: Concrete.
  Waterproofing: American Hydrotech hot rubberized asphalt in a protected membrane roof assembly.
  Green Roof System: American Hydrotech Extensive Garden Roof® Assembly.
  Drainage: Floradrain FD25 Drainage/Water Storage/Aeration component.
  Irrigation: None.
  Growing Medium: 5" depth.
  Vegetation: Fifteen species or varieties of Sedum are represented. (*See Appendix B: Plant Lists*)
  Design Objectives: New amenity spaces (recreation spaces); community involvement; storm water management: quantity and quality; energy efficiency; integration with other building components (Heating ventilation air conditioning units); waste diversion.

Like most green roofs, the one adorning the roof of the Eastern Village Cohousing Condominiums (EVCC) was born out of a highly collaborative design process involving a significant degree of input and direction from the owners. Cohousing is a type of housing in which residents actively participate in the design and operation of their own neighborhoods. It combines private home ownership with shared community facilities, activities, and decision-making. Operating within this model the future residents of EVCC (70 per cent of units were sold during the design phase) were engaged in all aspects of project development.

This intense involvement in the adaptive reuse of the abandoned 1957 office building served to shape it into the environmentally sustainable 56 unit urban community that it is today. The housing project is the first of its kind to be LEED® silver certified, incorporating numerous green technologies like low-flow water fixtures, high-efficiency lighting, and low-VOC finishes. Additionally, recycling during demolition and construction reduced project waste by 53 per cent. *(See Appendix D: Green Roofs and Leadership in Energy Efficiency Design)*

The green roof exemplifies the group's commitment to the environment, namely with respect to storm water runoff. The property is located in the Chesapeake Bay Watershed, the largest estuary in the United States and home to an incredibly complex ecosystem that includes important habitats and food webs. The Bay suffers from an excess of nutrients and contaminants washed into the water from impervious surfaces and agriculture.

Prior to construction, more than 90 per cent of the housing site was impervious and subject to occasional flooding problems. The green roof, the conversion of a parking area into a green space with native or adapted plants, and the construction of dry wells in the courtyard decreased the imperviousness of the site by more than 54 per cent and eliminated instances of flooding. These efforts were supplemented by rain barrels that collect runoff from the roof, redistributing it through irrigation of the at-grade landscape. These upgrades allowed the site to meet the Maryland Department of Environment requirements for storm water management.

On top of being a key environmental feature of the property, the green roof acts as a communal space for meetings, socializing, and relaxation; the west wing features extensive programming for residents. It is comprised of paved walkways, a gazebo, two pergolas, a children's play area, and rail-mounted planter boxes. When funding becomes available, it will host a meditation garden and raised vegetable beds. Likewise, the former location of rooftop mechanical units rendered unnecessary by ground-source heating and cooling has been set aside for a hot tub.

Though the green roof is evolving following the desires and capabilities of its users, it was initially designed to require minimal maintenance. The irrigation-free sedum planted extensive area requires annual weeding and slow-release fertilization. Much like the roof, the residents' green roof knowledge and maintenance strategies are maturing. They have brought in professionals, like Ed Snodgrass of Emory Knoll Farms, to further inform their maintenance practices.

Amid the rooftop plantings, cohousing members and visitors enjoy sweeping views of Silver Spring and Rock Creek Park. Members of the greater community have had several opportunities to partake in educational tours of the building and roof, which serves as an excellent model for the successful redevelopment of an unused office into a highly functional and environmentally friendly living space.

Residential Projects and Award Winners 49

Residents enjoy a meadow of sedum much as private homeowners would a lawn.

Sedum spurium in full color two years after installation.

Residential: Intensive

Seapointe Village Deck

Restoration

Location: Wildwood, New Jersey
Project Type: Retrofit
Green Roof Size: 4.5 acres
Completion Date: 2005
Year of Award: 2006

Client/Developer: Seapointe Village Master Association
Landscape Architect: Jeffrey L. Bruce & Company, LLC. *(Winner)*
Structural Engineer: Feld, Kaminetzky & Cohen, P.C.
Associate Landscape Architect: Edgewater Design LLC
Construction Manager: Gilbane Building Company
Civil Engineer: Barone Engineering Associates, LLC

Green Roof System Details
Structural Deck: Reinforced poured in place concrete.
Waterproofing: American Hydrotech, Inc. hot rubberized asphalt membrane in a protected membrane roof assembly.
Drainage: Thin composite drainage boards.
Irrigation: Conventional spray and low volume drip irrigation system controlled by volumetric water sensors.
Growing Medium: A custom blended formulation of sand, lightweight aggregate, and compost using locally available materials which resulted in the development on new material testing protocols.
Vegetation: Native salt tolerant materials that were adapted to the harsh coastal environment.
Design Objectives: New amenity spaces (recreation, horticultural therapy, food production); storm water management: quantity and quality; research applications; educational opportunities; ecological restoration.

Residential Projects and Award Winners 51

Two workman preparing the deck for upgrading.
Courtesy of: Jeffrey L. Bruce & Company

Set on seventeen oceanfront acres, Seapointe Village is considered to be a premier oceanfront resort community in Wildwood Crest, New Jersey. Three condominium towers and five and a half acres of recreational amenity space over a parking structure were constructed between 1986 and 1992. Following completion, the original 170,000 square foot green roof began to leak, damaging the concrete structures and vehicles below. Ten years of numerous unsuccessful repairs followed by a lawsuit led forensic experts to discover that the structure was subject to an accelerated rate of deterioration due to failed waterproofing in some areas and had a great potential for collapse.

Enter Jeffrey L. Bruce & Company and Edgewater Design, who were assigned the task of reconstructing the surface elements of the amenity space, while maintaining the integrity of the original landscape design and complying with new building codes and accessibility requirements of the *Americans with Disabilities Act* (ADA).

In order to correct the structural deterioration, the deck was reinforced with an additional four to eight inches of concrete. Designing this reinforcement was exceedingly complex, since the additional layer raised the roof deck six inches above the door thresholds of 130 units, meaning that water would drain into ground level apartments.

To avoid this situation, the topping slab was graded with a four-inch variation, sub-deck drainage and overflows. This upgrade allowed the roof to withstand the weight of the green roof and its users. However, it could not tolerate the live loads imposed by construction equipment during this delicate reconstruction. As a result, conveyers and motorized concrete buggies were used to place landscaping materials on the roof.

The roof system was designed with redundancy to minimize the reoccurrence of leaks. For example, the waterproofing membrane's thickness was doubled in high-risk areas such as water features. This precautionary attitude was extended to the implementation phase: the drainage layer and root barrier were installed over the membrane to prevent damage from other trades, decorative walkways, and plant material in planters.

Similarly, the system was designed to reduce the number of deck penetrations, minimizing the opportunity for problems caused by future concerns. The irrigation system distribution piping (comprised of advance water management features such as volumetric water sensors) as well as the control wires and electrical services were installed above the topping slab in a decorative concrete pavement.

To add to the project's complexity, the parking structure underneath the roof is located below sea level in a runoff restricted water-quality district. The drainage system was designed to local code storm water management requirements. The growing medium has percolation rates of six to fifteen inches per hour, resulting in little to no surface runoff. Furthermore it accepts and stores up to four inches of volumetric water for reuse in the green roof. When it reaches field capacity, the system releases the excess water into thin composite drainage boards and, following the slope of the deck, to roof drains.

The New Jersey shore is susceptible to airborne salt and hurricane strength winds that impose harsh conditions on the roof. To ensure the green roof's ability to withstand these environmental pressures, the designers applied emerging technologies such as lightweight growing media, contoured structural foam, and sand-based sod technology. Local native plants were selected for these same reasons and for their summer color, ability to minimize maintenance, conserve water, and enhance biodiversity.

The Seapointe Village experience points to the importance of the hidden components of green roofs and the associated technical challenge involved in attaining two seemingly contradictory goals: avoiding water leakage and retaining water on the roof's surface to support plant life. Had the outdoor space not been highly valued by the owner with his determination to restore this historic landscape, this project could have been a sad legacy for green roofs in New Jersey.

"Every project has its rough spots but this one was one of the most delightful projects we have ever worked on in terms of communication, documentation, and management. What truly made this project successful was the incredible working relationship between all design, construction management, owner management teams."
— Jeffrey Bruce, Landscape Architect,
Jeffrey L. Bruce & Company

Amenity spaces for residents' summer enjoyment. *Courtesy of: Jeffrey L. Bruce & Company*

Seapoint Village's completed north plaza. *Courtesy of: Jeffrey L. Bruce & Company*

Residential: Intensive

10th @ Hoyt Apartments

Location: Portland, Oregon
Project Type: New Construction
Green Roof Size: 8,500 square feet
Completion Date: 2005
Year of Award: 2006

Owner: Prometheus Real Estate Group
Client/Developer: Trammell Crow Residential
Landscape Architect: Koch Landscape Architecture *(Winner)*
Architect: Ankrom Moisan Associated Architects
Structural Engineer: Kramer Gehlen & Associates
Civil Engineer: David Evans & Associates
Mechanical Engineer: Interface Engineering

Green Roof System Details
   Structural Deck: Post tensioned concrete slab.
   Waterproofing: W.R. Grace & Company fluid applied membrane in a protected membrane roof assembly.
   Drainage: Hydrotech drainmat with 6 inches of drainrock in planted areas; cisterns.
   Irrigation: Conventional irrigation.
   Growing Medium: 10-30" of ProGro Mixes & Material.
   Vegetation: A variety of trees, shrubs, and ground covers. (*See Appendix B: Plant Lists*)
   Design Objectives: Storm water management: quantity; aesthetic improvement (integration with art objects); rain water harvesting; noise reduction; new amenity spaces (recreation spaces).

Storm water flowing over Cor-ten weir boxes and rocks.
Courtesy of: Koch Landscape Architecture

Located in a dense area of new development near downtown Portland, the 10th @ Hoyt Apartments courtyard provides semi-private respite for occupants and passers-by. The Persian inspired green roof, open to the street on the west side, was designed to capture, convey, and creatively display the storm water roof runoff. Water features, an important aspect of Persian landscape design, are integrated with plant material in raised and at-grade planters, which capture and filter additional rainwater. Seat walls and partially covered formal seating integrate the in- and outdoors and allow for rest and enjoyment of the courtyard.

The City of Portland regulations for on-site storm water mitigation require that owners of impermeable urban sites, such as this one, mitigate the rapid pace of storm water's entry into the public system. Many engineered solutions have been identified and implemented to satisfy this regulation, but visual display and celebration of this resource have seldom been implemented.

Three downspouts channel nearly all roof rainwater into pre-cast concrete channel systems – two of which empty water into shallow detention basins filled with decorative stones. The other sends water into a concrete cistern, which retains the water and then circulates it through a system of Cor-ten steel, usually used for ship building, weirs penetrated by glass buttons that are lit from within. All of these hard elements serve to create subtle textural contrast with the plant material, enhancing the garden experience.

After several hours, water is slowly released into the City's system, thereby reducing the size of the storm water interceptor from conventional sizing requirements. A pervious paving profile, planters, and the detention cistern perform additional retention and cleansing functions. The storm water detention elements, in conjunction with other lighting systems create a restful ambiance that exemplifies the many intrinsic auditory, visual, and tactile qualities of water.

The system has the capacity to hold all the roof rainwater for a one eighth of an inch storm event and detain rainwater for approximately thirty hours, relieving the first storm flow burden into the urban storm water interceptor system. As a result the courtyard has assumed, and exceeded, the required 20 per cent storm water mitigation function for the project. Koch Landscape Architecture conducted extensive flow studies and calculations to determine flow volume and characteristics. However, areas of the runnels created splashes and required that some of pieces be heightened. Since this is the first project of its kind, its lessons will inform future water based installations.

The limited construction timeline proved to be a challenge for the implementation on this very wet roof, especially in the case of the concrete cisterns. In cases like these, concrete usually requires forty-eight hours to cure and then it is coated with some form of waterproofing, which all told takes more time than the schedule would allow. Luckily, contractors bought used hydrophobic concrete that, once dry, was immediately ready for use.

Given its shady conditions, the courtyard will need very little irrigation to maintain its vitality. A simple landscape maintenance program and an annual cistern cleaning is all that is required to ensure that residents and the public continue to enjoy this innovative oasis that turns storm water into art. This courtyard may serve as a source of inspiration for landscape architects to innovate and explore new aesthetic dimensions in storm water intervention.

"This project acts as a precedent setting model for a well-integrated storm water system that is artful and dynamic. The multiple levels of color, texture and sound capture people's interest in the day and night."
— Steve Koch, Landscape Architect,
Koch Landscape Architecture

Vegetation at night, lit with Cor-ten weir boxes. *Courtesy of: Koch Landscape Architecture*

The formal seating areas integrate the in- and outdoors and allow for rest and enjoyment of the courtyard. *Courtesy of: Koch Landscape Architecture*

Residential: Semi-intensive

Lot 8 Santa Lucia Preserve

Location: Carmel, California
Project Type: New construction
Green Roof Size: 4250 square feet
Year of Award: 2007

Client/Developer: Dan & Sandy Feldman
Architect: Feldman Architecture
Landscape Architect: Blasen Landscape Architecture
Green Roof Consultant: Rana Creek *(Winner)*

Green Roof System Details
    Waterproofing: American Hydrotech, Inc. hot rubberized asphalt membrane in a protected membrane roof assembly.
    Drainage: Floradrain drainage board covered with 3/4" to 1/2" gravel with perforated pipe and surface drains at the roof's edges.
    Irrigation: Drip irrigation.
    Growing Medium: 6" depth of composed primarily of sand, lava rock, and amendments, such as mycelial inoculants.
    Vegetation: Perennial plant species from Oak Woodland and Monterey Peninsula region flora, complemented by annual wildflowers. (*See Appendix B: Plant Lists*)
    Design Objectives: Ecological restoration; visual integration with the surrounding environment; new amenity spaces (recreation); storm water management: quantity and quality; noise reduction; energy efficiency.

The green roof is adorned with native species from Oak Woodland and Monterey Peninsula. *Courtesy of: Rana Creek via. Green Roofs for Healthy Cities*

Built as an embodiment of Nature's gifts, the Feldman's home, in the Santa Lucia Preserve (a 20,000 acre private preserve in Carmel, California), was designed as an example of sustainable Mediterranean Modernism. The house was designed to integrate itself back into the land through ecologically sensitive features including low water use, solar power, and habitat enhancement. There are actually three small buildings that are built into the hillside where the hill seamlessly continues onto the green roofs of each one. Rana Creek ensured that an ecological design approach enhanced the project with a sustainable landscape and green roofs that consist mostly of locally adapted, indigenous plant materials already found thriving onsite prior to building. Their oversight of design and implementation focused on stabilizing all disturbed soils by planting grasses and forb mixes approved for the Santa Lucia Preserve, controlling non-native species, and simply allowing natural regeneration of the local plant assemblages. Adaptive management techniques proved to be the most valuable strategy, as the roof that was being "taken care of" by weeding, pruning, and irrigation was less successful than the roofs left unattended, due to lack of access.

The green roof design for the Feldman's Hill House emphasizes low water use, sustainable landscape techniques, and use of native plant materials. The green roofs are designed to provide usable landscape, filter and store rainwater, attenuate sound, increase thermal insulation, and provide site sensitive beauty for the Feldman's home. The Feldman's benefit by reducing their energy consumption up to 30 per cent during the summer months from the insulation of the green roofs. With a growing media depth of six inches the sound is reduced by approximately forty-three decibels. The sounds from the humans and their activities within the buildings are also being buffered to protect the wildlife, given the sensitive nature of the habitat in the Preserve.

The Feldman Residence is a superlative example of a green roof's potential to limit environmental disturbance and integrate architecture with ecology – the thirty-three species of native plants used have made for a successful recovery after the disturbance caused by building on the site. The ecology of the site will continue to become more complex and resemble the natural analogs that were emulated in the design!

"This project speaks to the Feldman's commitment to green design and their local environment."
— Paul Kephart, Restoration Ecologist, Rana Creek

Residential Projects and Award Winners

Feldman's Hill House emphasizes low water use, sustainable landscape techniques, and use of native plant materials. *Courtesy of: Rana Creek via Green Roofs for Healthy Cities*

Opposite: Neighboring buildings enjoy the view of the dynamic green roof. *Courtesy of: Gregg Galbraith Red Studio*

Residential: Semi-intensive

The Louisa

Location: Portland, Oregon
Project Type: New construction
Green Roof Size: 6,296 (extensive) and 8,071 (intensive) square feet
Completion Date: 2005
Year of Award: 2007

Client/Developer: Gerding Edlen Development
Landscape Architect: Walker Macy *(Winner)*
Architect: GBD Architects
General Contractor: Hoffman Construction

Green Roof System Details
Structural Deck: Post tension concrete.
Waterproofing: American Hydrotech, Inc hot rubberize asphalt membrane in a protected membrane roof assembly.
Drainage: Drainage mat with geotextile.
Irrigation: Automatic drip irrigation.
Growing Medium: 4" – 24" depth.
Vegetation: The extensive plot holds shallow root drought tolerant plants and the intensive plot holds native and drought tolerant plants. (*See Appendix B: Plant Lists*)
Design Objectives: New amenity spaces (recreation); building marketability; storm water management: quantity and quality; ecologically sensitive maintenance.

Residential Projects and Award Winners

Opposite: The green roof is strategically designed to maximize its enjoyment through public spaces and private gardens. *Courtesy of: Gregg Galbraith Red Studio*

The design of Louisa, a 242 unit residential high-rise in Portland, Oregon, was driven by ecological and social goals. These design objectives not only governed the materials used and the form they took, but also its design, implementation, and maintenance processes. The green roofs, one intensive and one extensive, are primary examples of the application of the principles of sustainability in every sense of the word.

Gerding Edlen Development and the design team, led by GBD Architects and Walker Macy, have worked together on several projects and, as a result, have established excellent lines of communication. These pre-existing relationships paved the way for a highly communicative and openly collaborative approach to the execution of the building and green roof.

Like many green roofs in the Portland area, storm water management was a key factor affecting material selection. The roofs are designed to store the minimal amount of water to keep the soil saturated, but because of the regional climatic pattern—extremely wet winters and extremely dry summers—the volume of water stored therein cannot equal the demand. Supplemental irrigation is needed in hotter months; it is used judiciously and only when evapotranspiration rates are at their highest. Designers also considered the implication of the green roof for the often-overlooked issue of storm water quality, specifying fertilizers that have no phosphorus content.

All maintenance procedures reflect the ecological principles that informed this building's conception. For example, the primary maintenance task is the manual removal of unwanted plants in lieu of the use of herbicides. Landscape architects and maintenance contractors continually monitor changes in growing medium composition, manage plant health, and irrigation balancing. These observations are used to inform the site's maintenance processes and design teams working on other green roof projects.

Since The Louisa was planned according to the LEED® silver certification criteria, the technical objectives for the project were established early in the design development phase. *(See Appendix D: Green Roofs and Leadership in Energy Efficiency Design)* This left the design team to focus greater attention on the social implications of design. They felt it was important for the green roofs to reach out to the surrounding neighbors by providing a stimulating tapestry of plant textures and patterns. The view is so coveted that units overlooking private gardens have paid a premium to have this daily dose of greenery.

Those lucky enough to live in the building have full access to the intensive green roof, allowing them to throw outdoor parties and barbeques. To further a sense of ownership of this green space, sections of it are portioned off as gardens for adjacent units. In order to maximize privacy for these gardens, the communal section was planned so as to steer its users away from the building entrance and nearby garden plots toward a seating area situated near the roof's periphery.

Fortunately, the general public can also access the green roofs through educational tours. Thus far, it has been visited by professionals working in sustainable design, soil sciences, horticulture, landscape architecture, and city planning. Similarly, a Green Resident Manual is distributed to new occupants when they move into the building in order to demystify the green roof and other sustainable technologies used in its construction.

The Louisa has grown two lush and flourishing green roof systems. They are emblematic of the design philosophy of the designers, owners, and residents and embody an ethic of sustainability.

"Understanding the program goals of a roof is critical to its success. In this case, careful social space planning was instrumental in creating a space with multiple uses that is beautiful to look at from inside and above. It is a special oasis in the city which can allow you to simply enjoy a paper in the morning."
— Mauricio Villarreal, Landscape Architect, Walker Macy

Residential Projects and Award Winners

The green roof is composed of drought tolerant and native species to minimize the water used throughout the year. *Courtesy of: Gregg Galbraith Red Studio*

Institutional Projects and Award Winners
Institutional: Intensive

Ducks Unlimited National Headquarters & Oak Hammock Marsh Interpretive Centre

Location: Oak Hammock Marsh, Manitoba
Project Type: New Construction
Green Roof Size: 28,190 square feet
Completion Date: 1992
Year of Award: 2003

Client/Developer: Ducks Unlimited Canada and Province of Manitoba
Project Architect: Number Ten Architectural Group *(Winner)*
Landscape Design: Hilderman Thomas Frank Cram & Associates
Structural Engineer: Crosier Kilgour and Partners
Mechanical Engineer: MCW Consultants
Electrical Engineer: AGE Engineering
Civil Engineers: UMA

Green Roof System Details
Structural Deck: Concrete.
Waterproofing: EPDM membrane, replaced with a 2-ply SBS Soprema System.
Root barrier: A geotextile fabric above the insulation.
Rodent Barrier: A galvanized wire mesh above the insulation.
Drainage: Two-leveled drains that provide drainage at the top of the growing medium and roof membrane levels.
Irrigation: Drip irrigation for the first two years and self sustaining after that date.
Growing medium: 4-15" (100-400 mm).
Vegetation: Prairie Grass and wild flowers; Kentucky blue grass mixture and native plants. (*See Appendix B: Plant Lists*)
Design Objectives: Ecological restoration; visual integration with the surrounding environment; integration with other building components (heating ventilation air conditioning units); educational opportunities; research applications; community involvement; new amenity spaces (recreation); ecologically sensitive maintenance.

Institutional Projects and Award Winners 69

The roof blends seamlessly into the distance when viewed from the observation decks. *Courtesy of: Number Ten Architectural Group*

Driven by ecological principles, this pioneering project at Ducks Unlimited National Headquarters & Oak Hammock Marsh Interpretive Centre was one of North America's first introductions to modern green roof technology. The office space, educational, and research facility was built with the intent to educate the public about native species and their habitat while minimizing the structure's impact on the surrounding natural community.

The site, located in an internationally recognized and restored prairie marshland, is home to thousands of migratory birds and visited by countless birding enthusiasts. The sensitive nature of the marshlands and the building's design objectives mandated that Number Ten Architectural Group and the design team approach the project with exceptional care. This resulted in the construction of a green roof and berming areas used to blend the structure and site, minimizing the visual impact of the building from a "bird's eye view" and rooftop observation areas. These landscaping elements in conjunction with other infrastructural elements, like a near elimination of cooling equipment, limited the structure's ecological impact and quelled environmental groups' initial opposition to development.

The design of the green roof was not without its challenges. Given that this project was implemented in the early days of North American green roofing, its design and installation involved a steep learning curve. At the time of implementation, one of the key areas in which there was a knowledge deficit was with regard to the application of waterproofing membranes for deeper soils and a total vegetated environment. The original loose laid base membrane was susceptible to movement during freeze-thaw cycles and was dislodged as the growing medium expanded and contracted. The membrane was later replaced when more appropriate and affordable fully adhered waterproofing technology was available in the marketplace.

Another challenge was presented by the natural cycle of the native grasses used on the roof, which must be periodically burnt in order to maintain a healthy grassland environment that rejuvenates itself. As a result, the roof deck was designed to accommodate a three-year to five-year burn cycle and all building systems were positioned to account for the smoke produced by these controlled fires. These fires have been carried out twice thus far and have not resulted in any problems for the building or its surrounding environment.

The roof continues to be a key part of the site's innovation and research program. Not only is it used by the Centre's 200,000 annual visitors, who observe and learn about the surrounding marshland and its inhabitants, but it is also used to further inform green roof planting methods through continual experimentation with various seed mixtures planted in isolated roof areas.

The success of this project can be measured not only by the quality of construction and satisfaction of the building's permanent occupants, but also by those occupying the roof and its surrounding area. Rooftop inhabitants include species such as piping plover and ground squirrels. Its sustainability is further confirmed by the fact that the quantity of wildlife nesting after construction is equal to that which occurred historically.

"The success of this project is due to a large multidisciplinary team working closely with a well-informed client looking to build something world class and truly sustainable."

— Bob Eastwood, Architect,
Number Ten Architectural Group

Institutional Projects and Award Winners

The Centre is integrated into the landscape with berms and green roofs. *Courtesy of: Number Ten Architectural Group*

Institutional: Semi-intensive

Peggy Notebaert Nature Museum

Location: Chicago, Illinois
Project Type: Retrofit
Green Roof Size: 2,400 square feet
Completion Date: 2002
Year of Award: 2003

Client/Developer: Peggy Notebaert Nature Museum
Landscape Architect: Conservation Design Forum *(Winner)*
Design Architect: Perkins and Will
Structural Engineer: CE Anderson and Associates
Landscape Contractor: Church Landscape, Inc.

Green Roof System Details
  Structural Deck: Limited capacity for additional support, ranging from 40-90 pounds per square foot.
  Waterproofing: Sika Sarnafil, Inc.'s 60 mil thermoplastic membrane in a conventional roof assembly.
  Green Roof System: RoofScapes Roofmeadow® Types **III:** *Savannah* and **IV:** *Meadow.*
  Drainage: Moisture management fabric, 2"-8" layer of gravel drainage material.
  Irrigation: Sub terrain drip irrigation system. Pond refill valve supplies the wetland with water. Solar panel supplies electricity to a trickle pump.
  Growing medium: Four profiles include a 2.5" wetland, a transitioning 4" extensive profile to a 6" semi-intensive profile, and an 8"-10" intensive profile.
  Vegetation: 80 species of native and hardy ornamental plants.
  Design Objectives: Storm water management: quantity and quality; visual integration with the surrounding environment; integration with other building components (green walls; rain water harvesting for irrigation); educational opportunities; research applications.

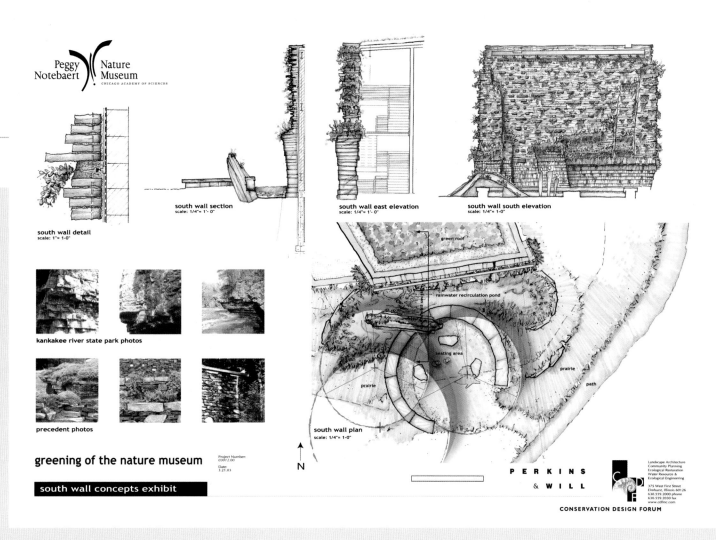

Every surface of the Nature Museum is designed to contribute to on-site storm water management.
Courtesy of: Conservation Design Forum

Opposite: This site demonstrates numerous green roof applications ranging from extensive to intensive to wetland applications. *Courtesy of: Conservation Design Forum*

The Peggy Notebaert Nature Museum project was conceived to publicly demonstrate the ecological benefits, system types, and design possibilities of green roof technology. This project was implemented in the early days of Chicago's green roof initiative. Having worked on Chicago's City Hall project, the designers at the Conservation Design Forum knew to exploit funding opportunities available through the Illinois Environmental Protection Agency and the U.S. Environmental Protection Agency through Section 319 of the *Clean Water Act*.

With funding secured, the designers proceeded to optimize the educational opportunities of the proposed roof area. However, the section of the roof allotted to this native plant green roof was one of great loading variation. Although it is directly visible from a number of vantage points within the museum, the limited loading capacity kept the project from being directly accessible to the public. At one end, the roof was capable of supporting an additional dead load of 90 pounds per square foot, while at the other it could only support 40. This constraint proved to be a blessing in disguise, forcing the designers to think innovatively about the optimal use of this 200-foot long space. Rather than apply the worst possible case scenario of the weaker end to the entire area they chose to create a gradual transition from a two and a half inch deep extensive application to an eighteen-inch deep wetland intensive system. This variation showcases the potential of this technology while providing habitat for a diversity of species.

Another factor that affected the organization and design of the green roof was the building's strong architectural geometry, which features dynamic sweeping rooflines and angular shaped masses of exhibit space. Combined with the long narrow shape, the green roof was in danger of becoming little more than a narrow strip of planting with a maintenance path leading straight to the north. The layout of the maintenance path however, became an organizing element to create harmony between the garden and the building. The dynamic zigzagged pathway divides the roof into smaller zones while respecting and complementing the lines of the building. The maintenance path provides a contrast between its hard angular lines and the informal arrangement of associated plants.

To help ensure the success of the green roof, it was designed to require minimal maintenance. The majority of the roof requires little maintenance, with the exception of the intensive section and the water element, which due to the slope has trouble maintaining the desired water levels.

The design of this green roof had implications for the site beyond that of most projects: it led to the development of a comprehensive master plan, in which every surface of the museum contributes to storm water management. The plan calls for the greening of every roof surface, save a small portion already designated for photovoltaic panels and a public terrace. Since this aspect of the project was completed, two other green roofs have been added to the building. Located over the main doorway to the museum and butterfly sanctuary, these plots demonstrate two other green roof system types that are suited to more arid conditions.

The master plan also mandates that all vertical surfaces be greened to blend the walls into the site in an organic way. Completed in the second phase of implementation, one of the green walls receives runoff from the roof and channels it to the North Pond, which then stores it for reuse. The Museum had already begun to incorporate native species into some of its at-grade plantings and Conservation Design Forum has taken this practice a step further, identifying potential areas for further expansion.

The Museum is committed to urban environmental education while enacting the principles that it promotes. The green roof has played a key role in raising public awareness about the overlapping relationship between the built environment and nature. In addition to providing educational opportunities, this project has allowed for plant survivability research and sets the stage for additional green roof work at the Museum and other buildings in the Chicago area.

"This project is an outstanding example of holistic integration of site ecology and building architecture in a way that is visible and educational to all visitors of the Museum."

— David Yocca, Landscape Architect, Conservation Design Forum

Award Winning Green Roof Designs

Peggy Notebaert Nature Center

The newer extensive roof provides a foreground for the Chicago Skyline.
Courtesy of: Conservation Design Forum

Institutional: Semi-intensive

The Church of Jesus Christ of Latter-day Saints Convention Center

Location: Salt Lake City, Utah
Project Type: New Construction
Green Roof Size: 1.1 million square feet
Completion Date: 2000
Year of Award: 2003

Client/Developer: The Church of Jesus Christ of Latter-Day Saints
Landscape Architect: Olin Partnership *(Winner)*
Architect: Zimmer Gunsel Frasca Partnership
Structural Engineers: KPFF
Theater Consultants: Auerbach + Associates

Green Roof System Details
Structural Deck: Concrete slab, with angled "bump" which crisscross the structure and support the growing medium.
Drainage: Sub-growing medium lateral perforated drainpipes, connected to surface drains, fan out from the structure's high point to the bottom of the roof; this pipe system is complemented by drainage board and aggregate.
Growing medium: Ranges from 2" in the meadow area to 4' depth in the treed area.
Vegetation: Varied native grasses and wildflowers, along with strategically placed *Pinus aristata* (Bristlecomb Pine). (*See Appendix B: Plant Lists*)
Design Objectives: Visual integration with the surrounding environment; storm water management: quantity; community involvement; new amenity spaces (recreation spaces); low maintenance.

The rooftop meadow was volunteer planted with grasses and wildflowers over the course of two weekends.
Courtesy of: Susan Weiler, Olin Partnership

Opposite: The 21,000 seat Convention Center was designed to mimic and blend into the surrounding mountain ranges. *Courtesy of: Marian Brenner, Olin Partnership*

In the late 1990s, the Church of Jesus Christ of Latter-Day Saints sought to build a center to house its conferences. The Church required that the center be large enough to accommodate 21,000 congregants and 1,400 vehicles without dwarfing the adjacent temple and surrounding landscape.

Through a deeply collaborative process, Olin Partnership and Zimmer Gunsel Frasca Partnership concluded that the best way to avoid creating an imposing structure would be to incorporate it into the surrounding landscape. This resulted in the creation of an eight-acre green roof supported by terraces and planted buttressed walls, whose vegetation mimics that of the surrounding Wasatch and Oquirrh mountain ranges. Taking advantage of a 65-foot change in elevation from one end of the site to the other, it was possible to integrate the roof, balcony, and orchestra levels of the auditorium with an extensive system of exterior stairs, terraces, gardens, and fountains.

The Church required that all auditorium seats have clear sightlines and open audio access while ensuring ease of entry and exit for all congregants. These internal design objectives eliminated the use of intermittent columns, restricting the depth of the green roof above the auditorium. Due to the careful attention paid to the structural limitations of large spans between roof supports the depth of growing medium varied from a minimal two inches in the meadow area to four feet to support fir and pine trees.

Collaboration did not end with the completion of the design phase, but wove its way through the entire development process. The client's support for the project was displayed through the assignment of several in-house Landscape Architects to the project during development, implementation, and maintenance periods. Their commitment was further demonstrated when the Church organized over 1,000 volunteers over the course of two weekends to hand plant the three-acre meadow. The community's contribution allowed for a more expensive planting method resulting in a shorter establishment period than would otherwise be possible.

The client's direct involvement in the design and implementation also resulted in on-going and knowledgeable care for the green roof. Involvement in all aspects of the project has allowed the Church to competently carry out modification of the green roof elements as needed, such as the replacement of some vegetation that proved to be ill-suited to the rooftop environment. Community dedication is also unwavering with members tending the site and mowing the meadow in the late fall. As the green roof matures its connection to the surrounding environment and community deepens, fostering social and natural communion while demonstrating new opportunities for greening Salt Lake City.

"This project speaks to the importance of collaboration between disciplines. It would not have been possible without the collaboration between and commitment of all individuals involved."

— Susan Weiler, Landscape Architect, Olin Partnership

Institutional Projects and Award Winners

The Church has dedicated landscape architects who monitor the roof and amend it as is necessary. *Courtesy of: Eckert & Eckert and Olin Partnership*

Institutional: Intensive

Opposite: A 'CloudGate' sculpture diffuses the light from the clerestory, one of several elements used to illuminate the building naturally. *Courtesy of: Roofscapes, Inc. Used by permission; all rights reserved.*

Oaklyn Branch, Evansville Vanderburgh Public Library

Location: Evansville, Indiana
Project Type: New Construction
Green Roof Size: 17,250 square feet
Completion Date: 2002
Year of Award: 2004

Client/Developer: Evansville Vanderburgh Public Library
Green Roof Designer: Roofscapes, Inc. *(Winner)*
Architect: Veasey Parrot Durkin & Shoulders
Landscape Architect: Storrow Kinsella Associates
Waterproofing Installer: Midland Engineering Company
Green Roof Installer: Enviroscapes

Green Roof System Details

Structural Deck: Steel/concrete composite roof deck.
Waterproofing: A Sarnafil, Inc. manufactured 80 mil PVC waterproofing, supplemented by an upper PVC protective membrane used to implement the base flood irrigation system in a conventional membrane assembly.
Drainage: 6" engineered drainage media.
Irrigation: Optigrün® base-level trickle irrigation system.
Green Roof System: Roofscapes, Roofmeadow® Type IV: *Meadow 1* assembly.
Growing Medium: 8" depth.
Vegetation: Native 'mesic meadow' prairie plant community bordered by informal group of red oak and flowering viburnum. (*See Appendix B: Plant Lists*)
Design Objectives: Energy efficiency; integration with other building components (elimination of internal drains); ecological restoration; educational opportunities; visual integration with the surrounding environment.

Conceived by Architect William B. Brown, a strong advocate for sustainable design, the Roofscapes, Inc. designed publicly accessible green roof is one of many innovative aspects of the Oaklyn Library. It was designed to require minimal maintenance, contribute to the structure's energy conservation, and utilize irrigation with minimal water loss through evaporation.

The design team was challenged by the Evansville Vanderburgh Public Library to link the roof to the local landscape. Thanks to the site's steep-sloping profile, the team designed an earth-sheltered structure. Contrary to popular belief this did not result in a dark and cave like interior, but rather in the most day-lit of all seven branches in the Evansville Library system. Light streams in through a dozen eleven-foot windows and the clerestory, in this case a band of windows perched above the roof. A 'CloudGate' sculpture diffuses the light from the clerestory and it rotates after hours to become a security gate, enabling use of the forward public meeting room.

When seen from the nearby highway, the mature roof blends seamlessly with the surrounding landscape; so much so that only the roof's dramatic 'Lightbridge' clerestory marks the library's presence. As the roof sweeps down to the building entrance it makes the transition from "naturalized" greenery to more traditional landscaping design. Informal groupings of red oak and flowering *viburnum* are arranged around the sloped building sides. Masonry parapet walls and a paved perimeter harmonize with the natural plant palette.

This direct connection with the surrounding environment resulted in the creation of a green roof that emulates the composition of surrounding prairies. The Mesic Meadow prairie plant community contributes to the ecological restoration of prairie landscapes in this region and provides a habitat for local creatures.

In further imitation of the surrounding natural system, characterized by a deep soil profile and perched water-table, the design team selected an irrigated two-layer Roofmeadow® Type IV: Meadow green roof system. This system, designed to last the life of the building, is comprised of a six-inch base layer of granular drainage medium covered by an eight-inch upper layer of growing medium. This profile has a Maximum Water Capacity (FLL) of four and three-quarter inches under fully drained conditions, or approximately 80 per cent annual storm water retention. Their installation marked the first time pneumatic blower trucks were utilized in North America to convey materials to the roof.

The roof deck was designed without roof drains to reduce the potential for long-term service and maintenance costs. Instead drainage is promoted by a roof deck pitched uniformly at 3 per cent toward one edge of the building. Normal runoff, reduced and slowed by percolation through the system, discharges from the down gradient side of the roof, where it is collected in a meandering grass swale. During large rainfall events, an internal drainage network composed of perforated rectangular conduits intercepts percolated water, directs and discharges it to a perforated storm sewer embedded in the swale.

Rising above the green roof, the roof and clerestory is a beacon to the community, a metaphor for the mission of the library visible for miles across the Pigeon Creek Valley. The library has become a vital community center with more than 35,000 patrons visiting each month, which is a three-fold increase in facility usage.

"Green roofs are only one component that can improve both the sustainability and beauty of a project. This project emphasizes how many ideas can be integrated gracefully in order to achieve measurable energy and environmental benefits."

— Charlie Miller, P. Eng., Roofscapes, Inc.

A direct connection with the surrounding environment resulted in the creation of a green roof that emulates the composition of surrounding prairies.
Courtesy of: Roofscapes, Inc. Used by permission; all rights reserved.

As the roof sweeps down to the building entrance it makes the transition from "naturalized" greenery to more traditional landscaping practice.
Courtesy of: Roofscapes, Inc. Used by permission; all rights reserved.

Institutional: Extensive

Opposite: A drought-tolerant plant suite, comprised of flowering sedums, was selected to create a dense and uniform low groundcover. *Courtesy of: Roofscapes, Inc. Used by permission; all rights reserved.*

Life Expressions Wellness Center

Location: Sugar Loaf, Pennsylvania
Project Type: New Construction
Green Roof Size: 6,000 square feet
Completion Date: 2001
Year of Award: 2004

Client/Developer: Ron and Joanne Gallagher
Green Roof Designer: Roofscapes, Inc. *(Winner)*
Architect: Van Der Ryn Architects
Green Roof Installer: David Bros. Landscape
Waterproofing Installer: Houck Services

Green Roof System Details
Structural Deck: Plywood on glulam rafters.
Waterproofing: Sarnafil G-476 reinforced 80 mil PVC waterproofing in a conventional roof assembly.
Drainage Layer: Fabric mat.
Irrigation: None.
Green Roof System: Adapted Roofmeadow® Type I: *Flower Carpet.*
Growing Medium: 5" depth, single layer.
Vegetation: 95 per cent of the drought tolerant plants are flowering *Sedum* varieties from the classic German "green carpet" suite, selected to create a dense and uniform groundcover. (*See Appendix B: Plant Lists*)
Design Objectives: Improved health and well-being (psychological improvements); visual integration with the surrounding environment; integration with other building components (elimination of internal drains); low maintenance.

Institutional Projects and Award Winners 87

Situated in sixteen acres of meadow in a mountain valley, the Life Expression Wellness Center was built to provide chiropractic services in a healthy, living environment. A green roof was mandated for the project by facility owners Ron and Joanne Gallagher, though project architect SimVan Der Ryn gave it its sinuous form. The owners wished to harness the spiritual effects of nature for their clientele, while minimizing the environmental impact of construction. Their interest in the technology extended well beyond the philosophical: they played an active role in every part of the design process, including plant selection. The thriving and well-maintained vegetation is a testament to their interest and dedication.

Serendipitously, Van Der Ryn and Charlie Miller of Roofscapes, Inc. met during the initial stages of design development and the two agreed to work together on what was to become one of the steepest green roof applications of its time in North America. The challenges posed by this roof included: stabilizing vegetation on the steep slope; protecting new plantings from severe mountain wind scour; and securing waterproofing at the overhanging fascia. Depending on the local slope conditions, material stabilization techniques included the use of roof battens, slope restraint panels, and reinforcing mesh. Due to weight and slope restrictions, leak detection could only be carried out by Electric Field Vector Mapping. This method can be used to locate leaks without disrupting landscaping components.

To further the therapeutic effects of natural phenomena, storm water runoff sheets from the eaves creating a curtain effect. This was accomplished by gapping the fascia by a half-inch, which required that Roofscapes, Inc. design unconventional waterproofing details suited to this unusual condition. The rainfall escapes through the fascia and then drips down tendrils of overhanging plants. The mature vegetated cover has slowed the rate and quantity of runoff prolonging the curtain effect.

The Gallaghers' also stipulated that the green roof vegetation be visually stimulating and easy to maintain. A drought-tolerant plant suite, comprised of flowering sedums, was selected to create a dense and uniform low groundcover. Because the vegetation is so dense weed pressure after establishment has been minimal. This resilience is also due to the xeric conditions on the roof.

Six years after completion, the site was surveyed by Roofscapes, Inc. to determine any variations in the system. In the early spring of 2007, vegetation had already achieved 90 per cent coverage, however the position of the plants was very telling. The peak and borders of the roof were very diverse, featuring over six species of sedum and some moss; whereas the highly sloped middle sections had but two, and an occasional three, sedum varieties.

The overall result of the green roof design has been to convert the building into a living structure that remains stable without active intervention. It is among a new generation of buildings that prioritize the health of both people and the environment.

"In cases like these, green roofs can make a powerful statement about the spiritual importance of nature in our living and working environment."
— Charlie Miller, P.Eng., Roofscapes, Inc.

Material stabilization techniques included the use of roof battens, slope restraint panels, and reinforcing mesh. *Courtesy of: Roofscapes, Inc. Used by permission; all rights reserved.*

The roof's vegetation and sinuous shape link the building to its surrounding environment. *Courtesy of: Roofscapes, Inc. Used by permission; all rights reserved.*

Institutional: Intensive

Opposite: The addition of vegetation and water features has contributed to patients' physical and psychological well-being. *Courtesy of: American Hydrotech, Inc.*

Schwab Rehabilitation Hospital

Location: Chicago, Illinois
Project Type: Retrofit
Green Roof Size: 10,000 square feet
Completion Date: 2003
Year of Award: 2005

Client/Developer: Schwab Rehabilitation Hospital
Landscape Architect: Douglas Hills Associates, Inc.
Architect: Stephen Rankin Associates
Roofing Contractor: E.W. Olson Roofing

Green Roof System Provider: American Hydrotech, Inc. *(Winner)*

Green Roof System Details
 Waterproofing: Upgraded the existing 5-year-old American Hydrotech, Inc. hot rubberized membrane.
 Drainage: Floradrain® 25 drainage board.
 Irrigation: Drip irrigation and selective water misters.
 Green Roof System: American Hydrotech, Inc. Garden Roof® Assembly.
 Growing Medium: 8"-18" depth (lightweight expanded clay aggregate, Canadian Sphagnum peat moss, compost, sand, and nutrient supplements).
 Vegetation: Early spring-blooming plants include Irises, Campanula, and Flax. In the summer, Daylilies, Cone Flowers, Russian Sage, Butterfly Bush, and Latria are in bloom. Fall selections include Aster as well as Sumac. Incorporated throughout these plantings are grasses such as Bunny Fountain and Little Blue Stem.
 Design Objectives: New amenity spaces (recreation spaces, horticultural therapy); improved health and well-being; aesthetic improvement (four season aesthetic); reduction of the urban heat island effect.

Institutional Projects and Award Winners

In 2001, Schwab Rehabilitation Hospital, a 125 bed comprehensive physical medicine and rehabilitation facility located on the West side of Chicago, wanted to provide its patients with access to a safe, barrier-free, and aesthetically pleasing environment. Given the limitation to the availability of space creating this type of environment on the roof made perfect sense.

When Schwab contracted landscape architecture firm Douglas Hills Associates, Inc. to prepare a concept master plan for the rooftop garden the hospital was in the final phases of a renovation project. This ultimately left little funding for the green roof's implementation. Thanks to the Landscape Architect's determination to see this project come to fruition, the firm discovered grant opportunities available through the City of Chicago's Department of the Environment's *Urban Heat Island Reduction Initiative*. This initiative, a precursor to Chicago's current suite of green roof programs, supports the utilization, development, and expansion of 'green' technologies that will mitigate the urban heat island effect and beautify Chicago. In total the hospital received a 400,000 dollar grant for its rooftop park and construction began in early 2003.

The midwestern style roof garden was designed to create seasonal interest and variety throughout the year. It is comprised of sweeping drifts of ornamental grasses and a small grove of ornamental shrubs that will remain beautiful and functional year round. Plants were selected for their fragrance, texture, and color to stimulate multiple senses. Other important features of the green roof design include a waterfall garden, a sixty-foot stream, and ornamental fencing on the existing parapet wall. In addition to being a restorative and beautiful space, the green roof was also designed to be a useful space, a place where the facility's healing programs could be carried on out of doors. The design team included Gene Rothart, Horticultural Services of the Chicago Botanical Gardens, who worked with Schwab staff to determine the type of horticultural therapy that would best suit this environment. Horticultural Therapy draws upon the culture of plant growth and gardening as a therapeutic tool for physical and mental rehabilitation. One of the key elements of the program was that the patients would be able to directly engage parts of the garden in raised planters beds, which would greatly increase the beds' accessibility. Initially, one of the programs also focused on food production on the roof.

The Hospital's original renovation roof team included American Hydrotech, Inc. and roofing contractor E.W. Olson Roofing, both with experience in green roofing. As a result, they were carried over to work on the new green roof. This utilization of experienced professionals resulted in a smooth construction process. However, after construction some growing medium erosion did occur, in which some areas slumped and piled over the parapet wall. The primary reason for this occurrence was that the gap between the top of the growing medium and the parapet was insufficient to accommodate the additional depth of dislodged growing medium.

The Schwab green roof is continually held up as a visionary aspect of this facility and has inspired a number of similar hospital projects. This intensive green roof provides an area for patients to exploit the healing powers of nature, drawing upon the culture of plant growth and gardening as a therapeutic tool for physical and mental rehabilitation.

"Horticultural Therapy is one of the least cited motivations for green roofs but it is one of the uses with some of the greatest benefits to the health and well being for end users. More application of this kind would be great."

— Steve Skinner, Project Manager, American Hydrotech, Inc.

Institutional Projects and Award Winners

This accessible green roof has been incorporated into the hospital's rehabilitation programming. *Courtesy of: American Hydrotech, Inc.*

In dense urban centers it is difficult to find at-grade land to create new, safe amenity spaces. *Courtesy of: American Hydrotech, Inc.*

Institutional: Extensive

The Evergreen State College Seminar II Building

Location: Olympia, Washington
Project Type: New Construction
Green Roof Size: 20,443 square feet
Completion Date: 2004
Year of Award: 2005

Client/Developer: The Evergreen State College
Architect: Mahlum Architects
General Contractor: DPR Construction, Inc.
Roofing Contractor: Wayne's Roofing
Landscape Contractor: Northwest Landscape
Green Roof System Provider: Garland Company *(Winner)*

Green Roof System Details
   Structural Deck: Cast-in-place concrete slab.
   Waterproofing: Garland's multi-ply SBS polymer modified bitumen roof.
   Drainage: Garland's Filterdrain 110.
   Irrigation: Drip irrigation, only used in times of drought.
   Growing Medium: 4-6" depth, Garland's Oasis media mixture.
   Vegetation: 33 low growing ground cover such as sedums and various perennial flowering species.
   Design Objectives: Storm water management: quantity and quality; reduction in onsite potable water use; integration with other building components (heating ventilation air conditioning units); energy efficiency; educational opportunities.

The green roofs were one aspect of a broader project goal to "celebrate rain" and respect the potential scarcity of water as a resource. *Courtesy of: The Garland Company, Inc.*

The Evergreen State College Seminar II building reflects the school's interdisciplinary teaching philosophy and commitment to environmentalism. This progressive College commissioned the structure in order to expand its teaching facilities while raising awareness about green building strategies and technologies. Perched atop the building are thirteen green roofs, only half of which are visible, that are emblematic of the college's green building mandate.

The green roofs were originally conceived as a way to bring organic food production to campus. Upon further investigation, the school's Building Committee and Mark Cork of Mahlum Architects determined that food production was not feasible on this project. However, the client recognized the many benefits of green roofs and insisted that they be a part of the project in one form or another.

This conviction that the inclusion of this technology was "the right thing to do" ended up proving true both environmentally and financially. Cork, at the suggestion of a county representative who was dissatisfied with the prescriptive method in which storm drainage systems were sized, commissioned a thirty-year computer model of the Thornton Creek watershed. This model showed the actual impact to the creek and demonstrated the additional storm water performance benefits of the green roofs. Though the model cost over 20,000 dollars to develop, its predictions resulted in a smaller stormwater detention system and resulting savings seven times the amount of the initial investment. Furthermore, the College now has a tool that it can use on any future developments.

The green roofs were one aspect of a broader project goal to "celebrate rain" and respect the potential scarcity of water as a resource. Water collected on the roofs is conveyed through surface landscape features to a central detention vault designed to release water back to the environment slowly. All onsite landscaping has a strict water usage policy: irrigation can only be used in times of drought. Similarly, the building's water fixtures are designed to reduce potable water use by at least 25 per cent.

In the same vein, the green roofs are tied into the building's energy conservation goals. The combined benefits of the reduced roof top temperature and the high r-value insulation provide the owner with a very energy efficient roof system. These benefits help reduce the energy needs of the building which was designed with no cooling system. In the few roof areas that could not be greened with one of 33 low growing ground covers or perennial flowering species, the project utilized Energy Star approved white coatings over the waterproofing membrane to further reduce energy consumption.

Many green roof installations are used as means to green building structures; this installation is not so much a part of the school's building as it is part of its spirit.

"The representatives of the College were very involved in the ideation and predesign processes. Their commitment to the project allowed for the construction of a building that was very sensitive to the nature of the surrounding environment."
— Mark Cork, Architect, Mahlum Architects

"It was inspiring to work on a project like this one and to see a building that is designed to provide long term solutions for a group of people with vision."
— Tom Regney, Garland Company

Institutional Projects and Award Winners 97

Low growing groundcovers cover thirteen rooftops.
Courtesy of: Mahlum Architects

All onsite landscaping has a strict water usage policy: irrigation can only be used in times of drought.
Courtesy of: Mahlum Architects

Institutional: Extensive

Ballard Branch of the Seattle Public Library

Location: Seattle, Washington
Project Type: New Construction
Green Roof Size: 20,500 square feet
Completion Date: 2005
Year of Award: 2006

Client/Developer: Seattle Public Library
Architect: Bohlin Cywinski Jackson
Landscape Architect: Swift & Co.
Green Roof Consultant: Rana Creek Habitat & Restoration
General Contractor: PCL Construction Services
Green Roof System Provider: American Hydrotech, Inc. *(Winner)*

Green Roof System Details
Waterproofing: American Hydrotech, Inc. hot rubberized asphalt in a protected membrane roof assembly.
Drainage: Drainage board.
Growing Medium: 4-6" depth.
Vegetation: Mixture of grasses and perennials species.
Design Objectives: Energy efficiency; integration with other building components (photovoltaic panels); storm water management: quantity; ecological restoration; educational opportunities.

The green roof stands in stark contrast with the very modern structure supporting it. *Courtesy of: American Hydrotech, Inc.*

The Ballard Branch of the Seattle Public Library was conceived to generate community interest in sustainable design by making the facility a dynamic teaching and research tool. The gently curving roof is visible to patrons through a periscope or from an observation deck and invites passersby to contemplate its aesthetic and ecological functions.

Self-sustaining, drought tolerant, indigenous grasses and sedums planted in a pattern that mimics a windborne casting of seeds stands in strong contrast with the stark structure below. Despite its wild appearance, the fescue and sedum plants were selected precisely because they would minimize the amount of maintenance the roof would require. To further simplify the installation and maintenance, the growing medium was blown on to the roof and held in place with biodegradable coconut fiber mats, reducing plant exposure to wind and water erosion during the two-year establishment period.

The faceted planes of the curved roof create six microclimate conditions. Each section is sloped and oriented differently, affecting exposure to the elements and water retention capabilities. The upper edges, those sloping more steeply, retain less water but offer the highest opportunity for evapotranspiration. On the other hand, the areas with lower slopes retain more water and are more protected from prevailing breezes. This diversity of conditions on one site provides an unprecedented opportunity for research on plant survival and evolution.

The building was also designed to minimize the use of potable water, employing a controlled irrigation system, low flow fixtures, sensor and timed faucets, and waterless urinals. Local groups have collaborated to install water-monitoring devices on the green roof. The data generated through this effort will be valuable in assessing the storm water performance of the green roof over the life of the structure.

The green roof was incorporated as part of an overall strategy to reduce and conserve energy costs where possible. Solar (photovoltaic) panels provided by the Seattle City Light Green Power Panel installed on the northern edge of the roof will be monitored for the amount of electricity captured and collected onsite. Energy generated from these panels is fed back in to the City's power grid, reducing the Library's energy bills. Additionally, various rooftop sensors measure wind speed, direction, and sunlight.

Ballard Library provides an excellent setting to model integrated high performance building design to the community by engaging them with highly visible features such as a green roof. The project not only demonstrates that a green building is feasible on a modest budget, it presents the community with an ideal example of the many benefits that may be realized when sustainable design combines with extraordinary architecture.

Monitored solar (photovoltaic) panels on the northern edge of the roof general electricity for the building below. *Courtesy of: American Hydrotech, Inc.*

The roof was carefully hand planted with a combination of grasses and perennial species. *Courtesy of: American Hydrotech, Inc.*

Opposite: The roof is located next to a light rail transit line and acts as a billboard for green roof technology. *Courtesy of: The Kestrel Design Group*

Institutional: Extensive

The Green Institute (Phillips Eco-enterprise Center)

Location: Minneapolis, Minnesota
Project Type: New construction
Green Roof Size: 4,000 square feet
Completion Date: 2003
Year of Award: 2006

Client/Developer: The Green Institute
Landscape Architect: The Kestrel Design Group, Inc. *(Winner)*
Project Architect: LHB Architects

Green Roof System Details
Structural Deck: Concrete.
Waterproofing: American Hydrotech, Inc. hot rubberized asphalt membrane in a protected membrane roof assembly.
Drainage: American Hydrotech, Inc. drainage board.
Irrigation: None.
Growing Medium: 2"-6" depth.
Vegetation: 11 European green roof species and 18 species native to Minnesota's Bedrock Bluff Prairies.
Design Objectives: Storm water management: quantity; ecological restoration; educational opportunities; research applications; new amenity spaces (recreation); community involvement.

Institutional Projects and Award Winners 103

The Green Institute, a Minneapolis based non-profit organization dedicated to sustaining the environment and local community through practical innovation, is home to the city's most frequently viewed extensive green roof. Located next to a light rail transit line, the roof is viewed by over 30,000 people a day.

To deepen the community's connection to the project, the institute organized several full day workshops about green roofing technologies and the installation process during implementation. One of the most important aspects of the workshops was that they served to debunk green roof myths and false expectations. For example, many volunteers thought that the planting period would be very quick, when in fact it took several days.

Now complete, the roof serves to educate visitors about the technical aspects of elevated planting and its interactions with the local ecological system. This physically accessible and high profile roof was designed to generate interest when inspected closely and glanced at from a distance. The roof is oriented following the four cardinal points, which are geometrically represented by the careful placement of various European Sedum species. Between these axes there are groupings of native wildflower and grass species from the Bedrock Bluff plant community. This layout locates the green roof within its surrounding social and biological context.

The Minnesota Bedrock Bluff Prairie plant community was used to inform native planting design, maximizing regional identity and ecological biodiversity, and minimizing maintenance requirements. Bedrock Bluff Prairies share many traits with traditional extensive green roofs in that they have shallow soil profiles and are exposed to considerable heat, drought, and wind. Members of this plant community are able to withstand these harsh conditions due to their thickened cuticles, hirsute stems and leaves, water storage cells, highly reflective surfaces, and fine or narrow leaves. The leaves' sticky surfaces can hold onto water while their leathery rough texture reduces the speed of surface wind. Instances of this plant community on the Mississippi River Bluffs have greatly decreased over the past few years. It is green roof designer Peter McDonagh's hope that green roofs will provide an analogous habitat within the Mississippi River watershed. Additionally, plant species were selected to provide a protected refuge for rare, threatened, and endangered moths, butterflies, and other invertebrates.

The Green Institute monitored the establishment rate and survival of eighteen native and eleven European species to determine the suitability of both for the creation of living, breathing buildings in Minnesota. Researchers have tracked the population of the native species since the project was completed and have noted a shift in composition. Some species have come to dominate areas of the roof while others have virtually disappeared.

In Minneapolis, the City of Lakes, storm water management is a regional driver for extensive green roofs. As a result, site researchers have also conducted research into the green roof's storm water management, quality improvement, volume reduction, and rate control. The green roof is capable of holding a one-inch twenty-four hour rain event: rains of one inch or less comprise 85 per cent of Minneapolis' annual precipitation.

The public education drive has paid off. Thanks to it and other initiatives, the green roof market in Minneapolis is growing at a rapid rate. There is now over 100,000 square feet of extensive greened rooftops in the Twin Cities and many more are on the way. Thanks to its high visibility and well-conceived design, The Green Institute's publicly accessible green roof continues to contribute to this phenomenon by providing an excellent opportunity for public education about the many benefits of this technology.

"This project has allowed us to examine the performance functions of different species in our local environment and has served to broaden the dialogue about plants in an extensive green roof application."
— Peter McDonagh, Landscape Architect, Kestral Design Group, Inc.

Over twenty-nine native and imported species were selected and are subject to comparison. *Courtesy of: The Kestrel Design Group*

Education about green roofs, green building technologies, and the surrounding natural communities were the driving forces behind this project's implementation. *Courtesy of: The Kestrel Design Group*

Institutional: Intensive

Mashantucket Pequot Museum and Research Center

Location: Manshantucket, Connecticut
Project Type: New construction
Green Roof Size: 65,000 square feet
Completion Date: 1998
Year of Award: 2006

Client/Developer: Mashantucket Pequot Tribal Nation *(Winner)*
Project Architect: Polshek and Partners

Green Roof System Details

Waterproofing: American Hydrotech, Inc. 120 mil hot rubberized asphalt membrane in a protected membrane roof assembly.
Drainage: Aggregate.
Irrigation: Timed drip irrigation.
Growing Medium: 12" depth composed of stratified layers of gravel, coarse/medium sands, and local topsoil saved from the construction site.
Vegetation: The original plant list included lowbush blueberry, daylily, liatris, strawberry, bloodroot, wormwood, yarrow, bee balm, spearmint, peppermint, black-eyed susan, wild columbine, tansy, and sage.
Design Objectives: Visual integration with the surrounding environment; storm water management; quantity; new amenity spaces (recreation and food production); educational opportunities; research applications; community involvement; ecologically sensitive maintenance.

Institutional Projects and Award Winners 107

Many of the plants used are culturally important and have traditionally been used as food, medicine, and raw materials. *Courtesy of: Bob Halloran, Mashentucket Pequot Museum and Research Center*

The years 1637-1638 saw the massacre and the near dissolution of the Pequot Tribe by European settlers at Great Cedar Swamp. Over the past three centuries the tribal lands have been subject to numerous land claim issues that were finally resolved on October 18th, 1983, through the enactment of the *Mashantucket Pequot Indian Land Claims Settlement Act*. This history of landlessness has profoundly affected members of the Mashantucket Pequot Tribal Nation's sense of place and their reverence for the land that they occupy.

As a result, when the tribe sought to build a museum and research facility, it wanted a structure that paid homage to the site's history while engaging the surrounding environment and maintaining the ecological integrity of the area. After much research and examination of European applications, a green roof was incorporated into the project as the primary vehicle for the realization of the project's environmental and educational goals.

Many of the plants incorporated into the terrace gardens are culturally important and subject to change as the roof evolves to suit the needs of the museum. Several "ethnobotany" gardens have been used to educate students about the cultural traditions of native people and their use of plants for food, medicine, and materials. The museum aims to create more gardens that reflect Pequot traditions. For example, museum curators aim to develop a root, herb, and berry garden that will supply the museum kitchen with periodic, local, seasonal, and traditionally harvested food.

Similarly, the function of the roof is subject to change as it is needed for different social functions. It has been used as an outdoor exhibit space, gathering area for museum receptions, demonstrations of traditional Native American games, and is accessible to the general public for meetings, lunch breaks, reading, and playing.

The roof's systems and maintenance processes further represent the tribe's relationship to Great Cedar Swamp. The roof was designed to minimize its water usage, employing drip irrigation on bushes and timed irrigation on each layer based on growing conditions. The irrigation cycles are determined using a potentiometer to measure water resistance in the growing medium. All excess storm water is channeled through ducts into a series of retention basins at the edge of the Cedar Swamp and allowed to rapidly filter back into the wetland. Detritus and contaminants in the runoff are minimal since maintenance staff compost grass clippings and use environmentally appropriate fertilizers.

The Mashantucket green roof is an excellent example of how this technology can serve as a place to preserve cultural traditions while at the same time upholding traditional beliefs in the sanctity of the land. "Hold on to the land" is a Pequot proverb born out of the land struggles. It is no surprise then that the museum project literally did just that, elevating the land from the ground to the roof.

"A notion of connectedness is embodied in the old Pequot expression 'Hold on to the land.' Not until recently has the tribe had the political and economic power to do so through initiatives like the green roof."
— Jason Mancini, Researcher, Mashantucket Pequot Museum and Research Center

Institutional Projects and Award Winners 109

The roof was designed to mitigate the impact of the site's stormwater runoff into the culturally important Great Cedar Swamp. *Courtesy of: Kirsten Orzechowski, Mashentucket Pequot Museum and Research Center*

The roof is frequently used as a gathering place for the community for cultural and educational purposes. *Courtesy of: Bob Halloran, Mashentucket Pequot Museum and Research Center*

Institutional: Intensive

Nashville Public Square

Location: Nashville, Tennessee
Project Type: Retrofit
Green Roof Size: 2.25 acres
Completion Date: 2005
Year of Award: 2007

Client/Developer: Metro Nashville Davidson County
Landscape Architect: Hawkins Partners, Inc. *(Winner)* and Wallace Roberts Todd
Architect: Tuck Hinton Architects
Civil Engineer: Barge Waggoner Sumner Cannon
Structural Engineer: Walker Parking

Green Roof System Details
Waterproofing: Tremco hot applied, rubberized asphalt waterproofing membrane
Drainage: Jdrain drainage board.
Irrigation: Automated conventional system.
Growing Medium: 8"-5' depth of Stalite growing medium.
Vegetation: 43 different species of which 81 per cent are native to the south-east region and 63 per cent are native to Middle Tennessee.
Design Objectives: Aesthetic improvement; new amenity spaces (recreation); visual integration with the surrounding environment; storm water management: quantity; reduction in onsite potable water use; ecologically sensitive maintenance.

Institutional Projects and Award Winners 111

The new park was designed to complement the renovated Davidson County Courthouse and Civic Building's 1930s art-deco architecture. *Courtesy of: Hawkins Partners, Inc.*

In 2003, the City of Nashville undertook a multi-million dollar renovation to its Metro Courthouse complex including a five level subterranean parking garage and "rooftop" public plaza. The total downtown site, consisting of approximately seven and a half acres, includes a two and a quarter acre state-of-the-art intensive green roof over the parking structure. A cornerstone of the design team's concept was the establishment of a truly civic open space that embodies the term "Public Square," providing unfettered access to all citizens to this civic hub from which a new pedestrian connectivity to the surrounding city could be realized. This new park not only complements the renovated 1930's art-deco styled courthouse architecture, but accentuates its grandeur from all perspectives using rich, timeless materials that are also authentic within its contemporary design interpretation.

This in every way is "context sensitive" design. Predominant civic views and axes literally shape the design's expression. At a more philosophical level, there is a subtle but important message sent by the broad civic lawn that stretches in front of the Davidson County Courthouse and Civic Building. From whatever point of entry, citizens reach the "level civic lawn" to stand equal in the sight of their elected government and court system.

Sensitivity to the stewardship of the environment, including water resources, is an integral part of the design. The challenge of a large impervious roof deck, the substrata of the park, is turned into an asset. Rainwater that falls on this area, as well as water from a garage sump, is harvested into a 57,000 gallon below-grade tank. Following filtration, this collected water supplies the required high-efficiency irrigation system with totally re-circulated water. Only in periods of drought is the tank topped-off with potable water. Due to this harvesting, it is necessary to employ an ecologically sensitive maintenance protocol that greatly reduces the use of chemical herbicides and pesticides — maintenance crews primarily utilize mechanical and organic weed and pest control.

The project re-incorporates many historically significant components from the original site. Some of those elements include reuse of granite units (wall caps, veneer, and steps), historical war commemorative monuments, and historical veteran plaques.

The Nashville Public Square design re-invents what once was a polluting surface parking lot, and with the resolve of its community leaders, shapes a civic gathering space of dignity and circumstance. It recalls and interprets the historic stories of this location. It accommodates all citizens, offering a barrier-free entry from the surrounding street grid. It turns the liability of an impermeable subterranean five-story parking garage into an asset that harvests rainwater, conserving hundreds of thousands of gallons of potable water each year, and employs green roof technology to make it possible.

"The Nashville Public Square reinvents what was once a polluting surface parking lot into a valuable civic gathering space. It turned the liability of a 2.25 acre subterranean 5 story parking garage into an asset that now harvests rainwater, increases the urban canopy, employees an intensive green roof system, and helps to heal the urban landscape."

— Gary Hawkins, Landscape Architect, Hawkins Partners, Inc.

Institutional Projects and Award Winners   113

The Nashville Public Square design reinvents what once was a polluting surface parking lot as a civic gathering space. *Courtesy of: Hawkins Partners, Inc.*

Rainwater that falls on this area, as well as water from a garage sump, is harvested into a fifty-seven thousand gallon below-grade tank. *Courtesy of: Hawkins Partners, Inc.*

Institutional: Extensive

Opposite: A drought tolerant green roof is connected to a storm water system that slows water outfall into Banklick Creek. *Courtesy of: Sanitation District N°1*

Sanitation District N°1

Location: Fort Wright, Kentucky
Project Type: New construction
Green Roof Size: 3,600 square feet
Completion Date: 2003
Year of Award: 2007

Client/Developer: Sanitation District N°1 of Northern Kentucky *(Winner)*
Architect: Humpert Wolnitzek Architects
Landscape Architect: Human Nature, Inc.
Green Roof Consultant: Roofscapes, Inc.
Landscape Contractor: Enviroscape

Green Roof System Details
Structural Deck: 3" metal roof deck.
Waterproofing: 80 mil EPDM with heat welded seams.
Drainage: 2" of granular drainage medium.
Irrigation: None.
Green Roof System: Roofscapes, Inc. Type III: Savannah.
Growing Medium: 2" depth.
Vegetation: Low maintenance drought resistant plants such as ornamental *Bouteloua gracilis* (grasses), *Allium schoenoprasum* (chives), and sedums species.
Design Objectives: Storm water management: quantity and quality; educational opportunities; research applications; rainwater harvesting for irrigation; low maintenance.

Institutional Projects and Award Winners

When Sanitation District N°1 took on the responsibility of managing storm water from thirty-two cities and three Northern Kentucky counties, it needed to expand the size and scope of its facilities. The District took advantage of this opportunity and built a facility that implemented the storm water best management practices (BMPs) it would be promoting, thereby demystifying technologies that are rarely used in the area. Events in the construction process served to reinforce the need for a local demonstration project: several individuals were reluctant to participate in the design process because they thought the design goals were unachievable.

A drought tolerant green roof was incorporated into the project and connected to a system of BMPs that would nurture storm water from the time it hit the roof to its outfall into Banklick Creek. The green roof is, in essence, at the headwaters of the entire site's hydrology with all the runoff flowing from the roof into a naturalized wetland through a retention basin, a detention basin and step pools, finally making its way into the creek.

These storm water management techniques are integrated into The District's child and adult educational programming. During the 2005-06 school year, the site hosted over 2,300 students and Scout groups to its environmental education site, Public Service Park. During a visit, the children play the role of scientists, tracking a drop of water through its ecosystem. This leads them through a wetland, a Native American Creek Trail, across porous pavement, and up to the roof. Interested groups, historically composed of professionals, civic groups and municipal bodies, can arrange for specialized tours of the roof and park. Furthermore, the sedum and ornamental grass roof is a local site available to university students interested in studying the application of green roofing technologies and other innovative storm water practices.

To further encourage the construction of green roofs in the area, the District sought to obtain performance data specific to the Northern Kentucky region. The green roof sits next to a conventional one that is being used as a baseline. Runoff from both roofs is collected and conveyed through a network of pipes to a storm water laboratory. In the lab, a series of parallel ten-inch clear PVC pipes retrofitted with water quality sampling instruments and flow monitoring devices enable dedicated research personnel to access qualitative (physical and chemical parameters) and quantitative (volume and rate of discharge) data during wet weather conditions. The collected data is currently being analyzed to determine the effectiveness of the green roof as a storm water management tool. Currently, they are measuring retention time, amount of storm water captured, rate of discharge, and potential water quality improvements. Data is collected year-round to determine the seasonal effectiveness of the green roof.

The Sanitation District N°1 project exemplifies the ideal use of an institutional green roof, not only embracing and exploring efficiency and utility possibilities, but research and education as well.

"By letting people experience the roof we are getting closer to achieving our on-site storm water management goal. Since this hands-on educational campaign began one green roof has been installed in the area and we continue to notice a growing interest in green roof applications."
— Jim Gibson, Director of Water Resources, Sanitation District N°1

The green roof is being used to generate region specific data. *Courtesy of: Sanitation District N°1*

Data is being collected from the green roof and an adjacent conventional roof to determine its effect on the quality and quantity of storm water runoff. *Courtesy of: Sanitation District N°1*

Industrial/Commercial Projects and Award Winners

Industrial/Commercial: Extensive

901 Cherry Street

Location: San Bruno, California
Project Type: New Construction
Green Roof Size: 69,000 square feet
Completion Date: 1997
Year of Award: 2003

Client/Developer: Gap Inc.
Architect of Record: Gensler & Associates
Design Architect: William McDonough + Partners *(Winner)*
Green Roof Consultant: Hargreaves and Associates, Rana Creek, and Ove Arup & Partners
Construction Manager: William Wilson and Associates
Green Roof Manufacturer: Sarnafil Roofing and Waterproofing Systems

Green Roof System Details
Waterproofing: American Hydrotech, Inc. hot rubberized asphalt membrane in a protected assembly.
Irrigation: Permanent system.
Growing medium: 6" depth.
Vegetation: Native grasses and wildflowers, re-establishes the indigenous coastal savannah ecosystem.
Design Objectives: Visual integration with the surrounding environment; ecological restoration; improved health and well-being; noise reduction; storm water management: quantity; rainwater harvesting for irrigation; ecologically sensitive maintenance.

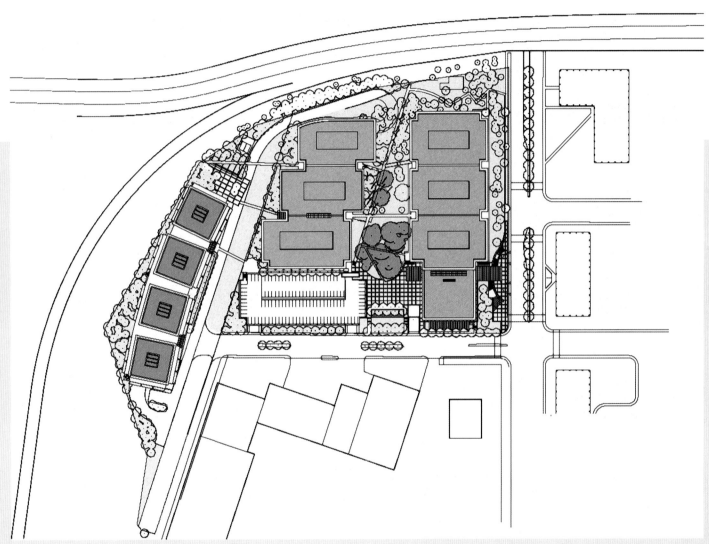

Seen from a bird's eye view, nothing has changed.
Courtesy of: William McDonough + Partners

Located in the hills above San Francisco International Airport, 901 Cherry Street blends almost seamlessly into its steeply sloping site; the undulating profile of its roofline mimics its hilly context. The building's connection to its location is more than skin deep: it is covered in native grasses and wildflowers, re-establishing the indigenous coastal savannah ecosystem it eliminated on the ground.

The green roof is the project's defining sustainable design feature. This single element had a fundamental effect upon the building's design profile, its physical relationship to surrounding context, its mechanical performance, acoustical and thermal comfort, and storm water management.

The design team sought to explore the notion of how a project can become native to its particular place, in this case, the savannah of the coastal foothills of the San Francisco Bay Area. That is: how can the design of the building be an act of atonement for the open space that the building is replacing? The roof is crucial to resolving this dilemma; to birds flying overhead, nothing has changed.

Despite its narrative elegance, however, measures had to be taken to gain the owner's approval at key junctures as the concept developed during the design process, up to and including the creation of a mock-up of the roof seal, gravel base, growing medium placement, drains, and vegetation. The schematic design process concluded with the simultaneous presentation of a detailed cost benefit analysis for all of the key green features on the project, including the green roof. Key functional criteria included the roof's impact on the structure, on roof longevity, and on storm water management, while extensive performance and aesthetic reviews provided further evidence of its suitability.

The thermal insulating characteristics of the proposed roof also offered annual energy savings of several thousands of dollars beyond that possible with a conventional roof assembly. Original projections, made years before the 2002 California energy crisis, showed that the green roof would achieve simple payback from enery savings alone in eleven years.

The roof's provision of additional thermal and acoustic insulation further integrated and consolidated its benefits. The roof's high thermal resistance — with a u-value three times better than that of a conventional roof — makes the building less susceptible to exterior temperature changes.

Combined with the building's unique and intensely site-specific use of nighttime air flushing, the insulating value of the soil allowed for the downsizing of heating and cooling equipment, contributing to overall energy savings. The mass of the roof also attenuates sound transmission by up to fifty decibels, no small consideration in a building in the flight path of a busy international airport. Of course, these benefits are further enhanced by the fact they will be enjoyed well into the future.

Aesthetically, the roof's primary goal is to mimic the appearance and ecology of the surrounding vegetation while allowing the architecture to integrate gently with the environment. The roof shape creates a graceful visual extension of the surrounding hills, furthered by the selection of native grasses and plants that reflect the natural environment's colors and textures. Grassland plants typically prefer gently sloping, well-drained soils, so the roof provides them with an ideal surface geography.

Like other native grasslands, the roof is a highly self-sustaining ecosystem, but as a part of the built environment, it provides a long-term, low-input landscape. Judicious plant selection from California's native palette of cool season grasses substantially reduced maintenance costs associated with mowing, irrigation, and fertilizers. Annual maintenance is limited to periodic irrigation during times of drought and mowing in advance of Fourth of July firecrackers.

The successful completion of 901 Cherry Street during the early days of the green roof industry has contributed to the growth and evolution of the industry.

> *"Blanketed in soil, flowers, and grasses, the roof's hills echo the local landscape, reestablishing several acres of the surrounding coastal savannah ecosystem. The roof also damps the sounds of jets from the San Francisco airport; it absorbs storm water, which is important because they have serious issues with storm water there; it makes oxygen, provides habitat; and, it is beautiful."*
>
> — William McDonough, Architect, William McDonough + Partners

Increased exposure to vegetation has been cited as a contributing factor to reduced employee turnover. *Courtesy of: William McDonough + Partners*

This green roof was designed to mimic the surrounding landscape and to mitigate the ecological impact of construction. *Courtesy of: William McDonough + Partners*

Industrial/Commercial: Extensive

Montgomery Business Park

Location: Baltimore, Maryland
Project Type: Retrofit
Green Roof Size: 30,000 square feet
Completion Date: Summer 2002
Year of Award: 2003

Client/Developer: Himmerlich and Associates
Architect of Record: Notari Associates
Design Architect: DMJM
Green Roof Consultant: Katrin Scholz-Barth Consulting, then with HOK Planning Group*(Winner)*
Roofing Consultant: Dedicated Roof and Hydro-Solutions
Structural Engineers: Morabito Consultants
Plant Supplier: Emory Knoll Farms, Green Roof Plants

Green Roof System Details
  Structural Deck: Tectum roof.
  Waterproofing: PVC Membrane in a protected roofing assembly membrane roof.
  Drainage: Internal gutter system.
  Irrigation: No permanent system.
  Growing medium: 3" depth.
  Vegetation: Variety of 17 drought tolerant sedum species.
  Design Objectives: Storm water management: quantity and quality; aesthetic improvement; building marketability; grey water harvesting; low maitenance.

Montgomery Business Park was renovated in order to enact environmental principles and implement the storm water best management practices that the Maryland Department of the Environment promote. *Courtesy of: Katrin Scholz-Barth*

When the Maryland Department of the Environment sought proposals for its new offices, Montgomery Park developers suggested a building strategy that would allow the department to implement its widely promoted sustainability agenda. The bid resulted in the large-scale adaptive reuse of the Montgomery Ward Catalogue Warehouse, originally built in 1925, using environmentally friendly and energy efficient technologies such as high-efficiency mechanical and electrical systems; ice storage; a gray water conservation system capturing storm water runoff; recycled building materials; insulated glass; and energy saving lighting with photocell dimming and occupancy sensors.

The most visible and easily identified "green" technology which has garnered the most attention for the building, is the first-story 30,000 square foot green roof adorning the former train shed. Occupants of the top seven floors of the horseshoe shaped building, as well as riders of glass-enclosed elevators, are granted a clear view of the greenery. Seen from above, this roof provides visual continuity between the building and the adjacent Carroll Park, community gardens, and golf course.

Though aesthetics was an important factor in the development of the green roof, its role in the on-site water quality strategy, as promoted by the Department of the Environment, was the key to its inclusion in the project. The green roof prevents storm water runoff and nutrient intake into the Gwynns Falls watershed, the Inner Harbor of Baltimore, and ultimately the Chesapeake Bay. During the design phase, a team of civil engineers surveyed the entire site's impervious area to estimate the reduction of storm water runoff associated with the inclusion of a green roof, a storm water cistern, and bioretention areas in the parking lot. Though the site was fully impervious, it was designed with a water retention capacity for a half-inch rain event. The storm water system captures all of the runoff from the parking area and green roof, which is then pumped up to a holding tank on the higher roof. The water is then utilized for toilet flushing in a gravity-based gray water distribution system.

Due to limited visibility and access and the presence of mechanical equipment on the higher roof, only the lower level roof was greened. This roof was structurally reinforced to accommodate the added green roof weight and two new twenty by forty foot skylights used to introduce natural light into the space below which now serves as a cafeteria, auditorium, and recreation center. These load changes required additional support, which was achieved by adding vertical beams every twenty-five feet to transfer the weight of the roof to the ground.

Other than this structural reinforcement, the design of the green roof was relatively straightforward. Due to the slight slope in the roof, no drainage layer was necessary. The growing medium and sedums were selected to satisfy the water retention requirements of the project while being able to withstand the arduous conditions on the roof.

Meeting deadlines and working in a tight schedule is always a challenge for complex renovations projects like this one. Despite the simplicity of the green roof system, its installation presented coordination and logistical challenges. One of the major issues was with regard to the protection of the exposed waterproofing membrane from activities by other trades. This resulted in the delay of subsequent implementation, requiring that rooftop vegetation be stored on an adjacent roof from April to June. Despite that year's hot and dry summer, 61,000 hardy plants survived and established quickly once they were installed, proving their applicability to the harsh rooftop environment.

The green roof has flourished and exemplifies the importance of green roof maintenance. The surrounding seven stories cast a shadow on areas of the green roof, which causes them to be moister than those sections continually exposed to the sun. Areas with elevated moisture levels are more conducive to the propagation of self-seeding weeds, which as of yet haven't created any aesthetic discord because they have been accepted as a natural outcome of the project.

"Within the Mid-Atlantic Region, this green roof project still has unmatched outreach potential and educational opportunity."
— *Katrin Scholz-Barth, Green Roof Consultant, Katrin Scholz Barth Consulting*

Industrial/Commercial Projects and Award Winners 125

Due to the slope of the roof, no drainage layer was required. *Courtesy of: Katrin Scholz-Barth*

The project employed many sedum species to minimize the need for maintenance and irrigation. *Courtesy of: Katrin Scholz-Barth*

Industrial/Commercial: Intensive

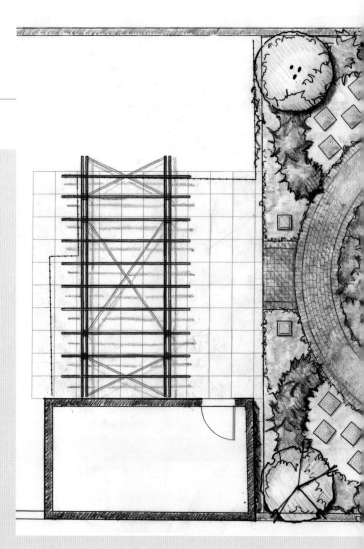

Garden Room

Location: Shorewood, Wisconsin
Project Type: Retrofit
Green Roof Size: 4,000 square feet
Completion Date: Summer 2002
Year of Award: 2003

Client/Developer: Garden Room
Architect of Record: John Schroeder
Landscape Architect: Buettner and Associates *(Winner)*
Landscape Contractor: Marek Landscaping
Roofing Contractor: Cudahy Roofing & Supply Inc.

Green Roof System Details
   Structural Deck: Concrete.
   Waterproofing: Siplast SBS Modified Bitumen.
   Drainage: Porous gravel layer with perforated PVC pipe covered with filter fabric.
   Irrigation: Milwaukee Lawn Sprinkler Corp.
   Growing medium: 18" depth.
   Vegetation: Variety of annuals, perennials, and shrubs. The four small ornamental trees are in insulated containers located directly over the roof support columns.
   Design Objectives: Aesthetic improvement; new amenity spaces (commercial).

Industrial/Commercial Projects and Award Winners

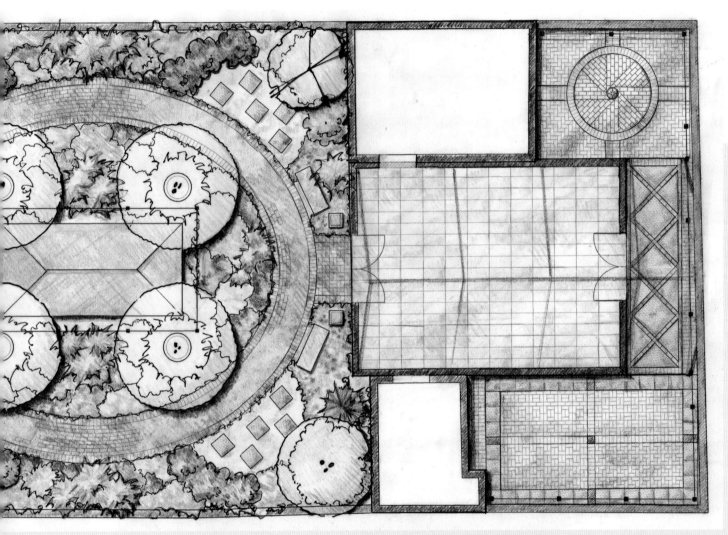

Courtesy of: Dennis Buettner, FASLA

The green roof on the Garden Room was designed to exploit the commercial and social value of greened roof space. A secondary goal was to educating customers about the construction, materials, function, and benefits of green roofs. Landscape Architect Dennis Buettner, in close collaboration with storeowner Deborah Kern, undertook the creation of three connected yet distinctly different spaces to achieve these goals.

The conservatory and two small brick patios are located at the front end of the building. This area used recycled brick street pavers dry laid in various configurations to raise consumer awareness about the diversity of brick paving patterns. The conservatory is adjacent to the intensively planted garden area. An oval brick path surrounds the skylight that illuminates the store's atrium below. Four seasons aesthetic interest was achieved using spring bulbs, summer blooming annuals and perennials, and woody plants with fall colors, winter fruit, and evergreen foliage. These compliment carefully positioned art, ornaments, and creative planting combinations in pots or planters. The garden terminates at the arbor-covered terrace used to display a variety of outdoor furniture. The arbor's patterned walls echo the oval of the walkway and hide unsightly mechanical and maintenance equipment.

The garden area proved to be that of the greatest technical difficulty for the design team. The fact that this elevated space replicated at-grade gardening exacerbated the common green roof problem of a limited structural loading capacity. The 12-18 inches of growing medium required to support intensive plants imposed a weight of 250 pounds per square foot: a weight comparable to that of two additional building stories. As a result, the wood deck supported by wall-to-wall 36-inch deep steel girders was replaced by a steel joist structural system with concrete on metal decking. This change further enhanced the green roof by allowing for a variation in roof pitch, which accommodated drainage requirements and varying growing medium depth.

Most importantly, the Garden Room has contributed to community revitalization. The roof, which was once atop a mechanic's shop, has become an evening gathering place for community members. Its public accessibility has also served to educate interested professionals in the greater Milwaukee area, hosting tours for groups like the Green Building Alliance. Furthermore, its visibility has served to increase rent and reduce turnover in apartment units overlooking the garden.

"I hope that people will look at this project as an example of how green roofs can be taken beyond utilitarian applications for stormwater management. The greater depth of growing medium supports a display a diverse plant material. The conservatory, garden, and terraces provide stimulating social spaces for evening use which benefits the surrounding urban neighborhood."
— Dennis Buettner, Landscape Architect,
Buettner and Associates

Industrial/Commercial Projects and Award Winners 129

Courtesy of: Dennis Buettner, FASLA

Courtesy of: Dennis Buettner, FASLA

Industrial/Commercial: Intensive

Burnham Park at North Soldier Field Redevelopment

Location: Chicago, Illinois
Project Type: Retrofit
Green Roof Size: 5.5 acres
Completion Date: 2003
Year of Award: 2004

Client/Developer: The Chicago Bears and The Chicago Park District
Architect: Lohan Caprile Goettsch Architects and Wood + Zapata
Project Manager: Hoffman Management Partners, LLC
Landscape Architect: Schaudt Landscape Architecture, Inc.
Irrigation, Soil and Turf Consultant: Jeffrey L. Bruce & Company *(Winner)*
Consulting Arborist: Urban Forest Management
General Contractor: Turner Construction Co., Barton Malow and Kenny Construction Company
Green Roof Installer: Monahan Landscaping

Green Roof System Details
Drainage: Aggregate drainage layer, or composite drainage board in shallower areas, leading to vitrified clay drainage pipe cut into the structural foam.
Irrigation: Large volume rotors to low volume drip, regulated by volumetric sensors.
Growing Medium: 6" to 30" "perched water system" profiles depth.
Vegetation: A variety of trees including caliper, maples, oaks, hawthorns, ash, and lindens; burning bush; winter creeper ground cover; and fine fescue turf grass.
Design Objectives: New amenity spaces (recreation); aesthetic improvement; visual integration with the surrounding environment; low maintenance.

Burnham Park's high use resulted in the development of many subsurface structural innovations. *Courtesy of: Jeffrey L. Bruce & Company*

The green roof of Soldier Field at North Burnham Park is part of a City of Chicago mandated parks project. The redevelopment of this one time landfill, containing debris from the Chicago Freight and Tunnel excavations, included the addition of two new parking garages and the renovation of the Soldier Field Stadium, home to the Chicago Bears. This project is set apart from others because of its technical complexity and by the tight timeline that it followed: the entire site's design and construction reclamation was carried out in a mere eighteen months.

Mayor Daley requested that the landscaping, though constructed in a very short period of time, mimic that of a mature park setting. This resulted in the installation of hundreds of large caliper trees, many of which were located on the roof deck. The plant palette was selected to blend in with the surrounding landscape and included maples, oaks, hawthorns, ash, and linden trees. Those parts of the green roof that were unable to support trees were planted with smaller plant material such as burning bush, winter creeper groundcover, and fine fescue, a low maintenance grass.

The degree and frequency of public access on the roof also posed challenges to the sustainability of the landscape. The arduous physical and loading abuse that the components would be subjected to required that consultants from Jeffrey L. Bruce & Company approach the agronomic design innovatively. Innovations included the use of contoured structural foam to increase depth in areas with limited structural loading capacity, friction mats and open celled structural matrixes for securing growing medium to slopes, turf reinforcement fibers for stability and structural bearing capacity in high use areas, non-blinding open matrix fabrics for erosion control, and washed sod technology. Furthermore, new soil laboratory testing protocols were designed to evaluate and understand specific agronomic performance of these new systems.

All growing medium profiles, of which there were nine, were tailored to maximize water efficiency and agronomic performance. Those on the roof deck are internally drained with percolation rates between six and fifteen inches per hour resulting in little to no surface runoff. Other profiles used on site were designed to perch water above an aggregate drainage layer similar to that within a golf green. When profiles reach material field capacity, water is flushed into the aggregate drainage layer and, using the slope of the deck, finds its way into a traditional vitrified clay drainage pipe cut into structural foam. Thin composite drainage boards, rather than drainage aggregates, were installed in those areas with shallow profile, maximizing water management by storing up to three inches of water.

Unfortunately, implementation of some of these newly engineered systems resulted in some unique challenges. For example a contractor was not able to achieve the specified compaction required to stabilize the slope using traditional landscape construction methods, despite having tried several methods. This necessitated that high slopes be redesigned to facilitate the local construction trade's ability to complete the task. Similarly, the parking structure's limited loading capacity could not support traditional construction equipment, resulting in the use of conveyors to place all growing medium and construction materials. This added a unique logistical complexity to an already technically challenging project.

Given the initial hesitation to embrace this publicly funded project, it has proven to be one of the most functional urban spaces in Chicago. The reinvigoration of this area and of the stadium structure speaks to the City's commitment to integrate greenery into Chicago's fabric.

"Burnham Park is a tribute to a forward thinking city that has extended its environmental ethic to its athletics field. While most municipalities are abandoning or demolishing old stadiums, Chicago chose to retain the historical components of this structure and renovate it using green technologies."
— Jeffrey Bruce, Landscape Architect,
Jeffrey L. Bruce & Company

Burnham Park replaces a one-time landfill and sits over two new parking garages adjacent to Soldier Field Stadium. *Courtesy of: Jeffrey L. Bruce & Company*

Contractors on the project were confronted with very steep and challenging slopes. *Courtesy of: Jeffrey L. Bruce & Company*

Industrial/Commercial: Extensive

Ford Rouge Dearborn Truck Plant

Location: Dearborn, Michigan
Project Type: New building
Green Roof Size: 454,000 square feet (10.4 acres)
Completion Date: Fall 2002
Year of Award: 2004

Client/Developer: Ford Motor Company
Concept Design Architect: William McDonough + Partners *(Winner)*
Architect/Engineer of Record: ARCADIS *(Winner)*
Design Consultant: McDonough Braungart Design Chemistry
Executive Architect: Giffels Inc.
Construction Manager: Walbridge Aldinger
Research Support: Michigan State University, Department of Crop & Soil Science and Department of Horticulture
U.S. Green Roof Consultant: Xero Flor America
Storm water Consultant: Cahill Associates, Arcadis Giffels
Roof Membrane Installer: ChristenDETROIT
Vegetation Consultant: Wildtype Native Plants
Vegetation Suppliers: Hortech, Inc. and Walters Gardens, Inc.

Green Roof System Details
Structural Deck: Metal deck with DensDeck protection board with 50 foot structural spans.
Waterproofing: Siplast SBS-modified bitumen membrane in a protected membrane assembly.
Root Barrier: Xero Flor 20-mil high-density polyethylene sheet.
Drainage: Colbond manufactured 1" (2 cm) thick nylon mesh bonded to a geotextile fabric.
Irrigation: Permanent irrigation monitored by a Rain Bird system, only used for initial vegetation installation.
Vegetative Mats: Xero Flor.
Moisture Retention Fleece: Mineral wool.
Growing Medium: 1" (2.54 cm) depth.
Vegetation: A variety of sedum species. *(See Appendix B: Plant Lists)*
Design Objectives: Storm water management: quantity and quality; ecological restoration; educational opportunities; waste diversion; ecologically sensitive maintenance.

Industrial/Commercial Projects and Award Winners 135

Covering 10.4 acres, the Ford Truck Plant green roof is the largest extensive green roof in North America.
Courtesy of: Ford Photography via. William McDonough + Partners

William McDonough + Partners has become synonymous with environmentally responsive design. It is no surprise then that Ford Motor Company would turn to this architectural firm in order to revitalize the historic Ford Rouge Center complex. In 1999, the complex and its surrounding environment had been severely degraded due to a near-century of industrial manufacturing. The building site consisted of rail lines and parking lots: it was completely denuded of vegetation.

Bill Ford's background in ecology allowed him to see the merits of a proposed ten and a half acre green roof on the new truck plant as a part of the site's greater storm water management plan. At the time, Ford Motor Company identified a savings of tens of millions of dollars because they were able to avoid constructing a water treatment facility mandated by proposed state Environmental Protection Agency regulations. Rainwater that is not retained by the green roof travels through a series of swales and wetland ponds where it undergoes natural treatment before returning to the Rouge River.

At the project's outset, the roof proved to be the source of much controversy with regard to the technology's performance and cost. It was made clear that auto production could not be interrupted due to potential roof leaks. In order to convince dubious members of the company, Ford staff and William McDonough + Partners' Roger Schickedantz worked with a team of scientists at Michigan State University to research different green roof technologies. As time progressed the idea began to gain momentum and staff began to buy into the concept.

The sheer size of the roof had several implications for design. Fifty-foot spans between supports severely limited the structural loading capacity of the roof, causing the design team to select a lightweight vegetative mat system. This system was well suited to the project not only because of its weight, but also its simplicity. A quick and straightforward installation process was needed due to competition for staging space with adjacent construction.

The one-inch pre-vegetated mats required a large nursery space for their manufacture. Fortunately, Ford owned a nearby landfill site and was able to create fields, tended by Xero Flor personnel, dedicated to the mats' cultivation. This move ensured that the right quantity of plant material would be available when needed and also reduced the potential for additional costs associated with plant storage caused by delays in construction. Nurseries usually charge a premium for storage past the established delivery date; this volume of plant material would have caused the price of holding stock to be exorbitant.

Ford placed orders a year in advance of the intended installation, so that Sedum cuttings could be cultivated during a full growing season. Vegetation was grown on the ground for over twelve weeks, after which it, and supporting geotextiles, were cut into one by one meter pieces, palletized, and transported to the site by truck and to the roof by crane. During the first year of plant establishment, sedums were fertilized using Rosasoil™, an organic liquid fertilizer derived from fermented vegetation that was delivered through a temporary irrigation system. A control release fertilizer was used for the rest of the establishment period. Since then, no fertilizer or irrigation has been used on the roof because over-watering will stress the sedum. Initially there were some areas with thinning vegetation, but over time they have been filled in with a dense carpet of greenery.

The Ford Motor Company wanted to keep maintenance and associated costs to a minimum. The size of the roof necessitated that it be subjected to continual maintenance, however the degree to which that was required would have considerable impact on the project's feasibility and sustainability. An initial proposal called for a plant pallet constituted primarily of native grasses. However, research from Michigan State University showed that unwanted plants and larger vegetation would only take root in growing medium deeper than two inches and would be unable to tolerate the xeric conditions and die off. Since a growing medium at that depth would greatly reduce the quantity of maintenance required on the project, designs were changed to accommodate species that could tolerate these conditions, i.e., sedums.

Even large industrial sites and their surrounding areas can benefit significantly from well-designed green roof infrastructure. This roof is the result of an approach that prioritized the performance functions of the roof, rather than its final form. It is truly a balance of ecology and economy.

"This project is a prime example of balancing economy, ecology, and equity. What we initially thought was the most ecologically correct design solution – native grasses – turned out to be the most difficult to maintain and the least economically feasible. At this scale, it would not have been sustainable."

— Roger Schickedantz, Architect, William McDonough + Partners

Vegetated mats were cut into one by one meter squares for transportation to the roof. *Courtesy of: Ford Photography via. William McDonough + Partners*

Tour groups can view the green roof from a nearby observation deck. *Courtesy of: Ford Photography via. William McDonough + Partners*

Opposite: The Lurie Garden refers to Chicago's transformation from its flat and marshy origins to a bold and powerful city. *Courtesy of: Terry Guen Design Associates*

Industrial/Commercial: Intensive

Millennium Park

Location: Chicago, Illinois
Project Type: New Construction
Green Roof Size: 24.5 acres
Completion Date: Summer 2004
Year of Award: 2005

Client/Developer: City of Chicago
Project Management: URS Corporation
Project Landscape Architect: Terry Guen Design Associates, Inc. *(Winner)*
Landscape Architect: Carol JH Yetken
Soil Scientist: Christopher B. Burke Engineering, Ltd.
Irrigation, Soil and Turf Consultant: Jeffrey L. Bruce & Company

Green Roof System Details
Structural Deck: Reinforced concrete cast-in-place garages.
Waterproofing: Henry hot-applied rubberized membrane and root barrier in protected membrane roof assembly.
Drainage: Sand layer which carries water to central drain pipes.
Irrigation: Centralized pump operating six irrigation zones, each section is irrigated as needed.
Growing Medium: 8"- 4' depth.
Vegetation: Over 900 trees planted with over 30 different varieties of deciduous and evergreen trees.
Design Objectives: New amenity spaces (recreation); aesthetic improvement (integration of art objects); improved health and well-being; community involvement.

Millennium Park was designed as a free cultural venue with a focus on providing a new state-of-the-art outdoor music venue in downtown Chicago. The design of the new park has transformed the physical and economic vitality of this foremost City location, turning rail yards and roadways into a world-class amenity space. While matching the patterns of the historic Michigan Avenue frontage, the park comfortably integrates cutting-edge technology and art into a multilevel contemporary usable public space. The roof top park holds many now renowned works of architecture, fountains, sculpture, and botanic garden spaces, as well as grand performance facilities, restaurants, and a skating rink. The green roof covers two new subterranean parking garages with 4,000 spaces, a multi-modal transit center including a bridge over the existing railroad lines and station, and a 1,525 seat indoor performance theater.

Having been subject to much conjecture since the 1970s, at last the site's final form was conceived by the Mayor's Office in 1997. The multimillion-dollar funding strategy involved constructing an underground parking facility whose revenue would support bonds to build a portion of the project and creating Millennium Park, Inc., a private donor group that would fund a dozen works of art in the park.

The Park consists of numerous tree-lined corridors leading to multiple "rooms," most notably the Lurie Garden, the Great Lawn and outdoor orchestra shell, and the Pritzker Pavilion. Terry Guen Design Associates, Inc. sought to unify all of these disparate elements through the creation of a common landscape language. This language was developed in collaboration with each artist to ensure that the greater context complemented the artwork, while enhancing the space's function. For example, the vegetation around the Gerhy designed BP Bridge acts as textural wallpaper on its reflective surface; its absence around Anish Kapour's Cloud Gate allows for the visual distortion of the city- and cloudscape to take place.

This project's size not only posed aesthetic challenges, but logistical ones as well. One instrument used to overcome the magnitude of the task was the use of recipe-like specifications written as a means to avoid many common implementation problems. They clearly communicated the design intent and installation methods to the hundreds of contractors working on site. This step help ensure that materials were installed according to green roofing best management practices throughout the Park, regardless of who was carrying out the task.

The division of the Park into "rooms" facilitated the implementation process by clearly delineating sectional responsibilities. To further simplify construction management, the site was halved into two major sections: East and West. Each half had a separate general contractor and operated on different timetables with staggered completion dates.

The challenges posed by its size did not cease with the Park's opening. With over one and a half million visitors since it's opening in 2004, the general public loves the park to death. As a result, strict maintenance protocols were created to accommodate the highly stressed natural systems. Advanced planning has allowed Millennium Park to embody a new type of park landscape, activated rather than bucolic, reflecting the lifestyle and interests of our current culture. The multiple uses of this unique space, both above and below the roof deck, are a testament to the potential of innovative city building through green roof infrastructure.

"We need to be demanding that our landscape perform within an environmental context and serve several functions. Millennium Park is a prime example of the way in which implementation and maintenance planning is a key component of its ability to satisfy its environmental, social, and cultural purposes."

— Terry Guen, Landscape Architect,
Terry Guen Design Associates, Inc.

Industrial/Commercial Projects and Award Winners

Frank Gehry's Jay Pritzker Pavilion and Anish Kapoor's Cloud Gate are two of the Park's most recognizable features, allowing visitors to listen to outdoor concerts, and reflect upon the city. *Courtesy of: Terry Guen Design Associates*

Jaume Plensa's Crown Fountain is a tribute to Chicagoans and features the faces of 1,000 residents. *Courtesy of: Terry Guen Design Associates*

Opposite: The colour pallet of the roof is constantly changing as different species bloom. *Courtesy of: Roofscapes, Inc. Used by permission; all rights reserved.*

Industrial/Commercial: Extensive

Heinz 57 Center

Location: Pittsburgh, Pennsylvania
Project Type: Retrofit
Green Roof Size: 12,000 square feet
Completion Date: Fall 2001
Year of Award: 2005

Client/Developer: The Huntley Group
Architect: Burt Hill Kosar Rittlemann Associates
Green Roof Provider: Roofscapes, Inc. *(Winner)*
Green Roof Installer: Lichtenfels Nursery
Waterproofing Installer: Burns & Scalo Co.

Green Roof System Details
 Structural Deck: Structural concrete.
 Waterproofing: Carlisle Syntech 045 EPDM waterproofing, supplemented by a polyethylene root barrier.
 Drainage: 2" drainage aggregate.
 Irrigation: None.
 Green Roof System: RoofScapes Roofmeadow® Type III: *Savannah* assembly.
 Growing Medium: 3" depth.
 Vegetation: 31 xeric species from 19 genera, including 6 North American natives – approximately 1/3 *sedums*, and a variety of herbs, meadow grasses, and meadow perennials.
 Design Objectives: Aesthetic improvement; new amenity spaces (recreation); low maintenance; noise attention.

Industrial/Commercial Projects and Award Winners

As part of an effort to reinvigorate Pittsburgh's urban core, The Huntley Group of developers retained Burt Hill Kosar Rittelman Associates to rejuvenate the former Gimbel's department store. The fourteen-story structure was originally built the 1930s and features a wealth of the period's classical detailing and a visual landmark in the city's populous business district.

Renovation plans called for a dramatic change that enticed the Heinz Company to locate their North American headquarters in the building's top seven floors. They were enticed by the insertion of a fifty-foot-diameter octagonal atrium in the middle of the structure, reaching from the fourteenth to the seventh floors. This one-acre floor plate introduced vast quantities of natural light into the upper stories of the building.

Late in the process, the architect and developer agreed upon the replacement of a proposed paved patio area surrounding the fourteenth floor penthouse offices with a green roof. The greenery would provide enjoyable views of a variety of flowering plants from the offices while hardscape areas could be used for outdoor meetings and gatherings. The ground cover would also provide an acoustical damper in the brick- and glass-enclosed terraces.

The late introduction of this part of the project required that the design and installation happen according to a very tight schedule. Roofscapes, Inc. was then brought in to help develop plans for the detailing and implementation. The time constraint imposed a very close relationship between the client and design team who elected to use a two-layer Roofmeadow® Type III: *Savannah* assembly. This system utilizes three inches of engineered growing medium over two inches of engineered drainage medium, in order to enhance the drought-resistance of the green roof ecosystem. It was then complemented by an undulating rhythm of planting around the roof perimeter designed in cooperation with Burt Hill's landscaping department. To capitalize on the changeable micro-climatic conditions created by high parapet walls, plants with diverse heights, textures, and bloom colors that would also tolerate the roof's xeric conditions were selected.

Accessible patio areas were constructed using high-density recycled plastic lumber roof decks. They are integrated with paved patios, which encourage the free flow of water between paved and vegetated areas.

The fact that the top floors were already occupied during implementation greatly increased the logistical complexity of the project from that on a vacant site. In order to avoid disturbing the new tenants, green roof materials were lifted to the fourteenth floor and then hoisted across the penthouse roof. Four months after its inception, the green roof was ready for use by the building's tenants.

Just like Gimbel's department store, the Heinz 57 Center's green roof is a highly visible part of the city's urban fabric. It demonstrates an effective means to improve the habitability of urban office space in downtown Pittsburgh. Citing this "flagship" example, the Mayor has been encouraged to introduce incentives for more urban green roof projects in Pittsburgh.

"It is never too late for a good idea. In many projects green roofs can be successfully introduced into design very late in the construction design process."
— Charlie Miller, P. Eng, Roofscapes, Inc.

High parapet walls allowed for plants with diverse heights, textures, and bloom colors. *Courtesy of: Roofscapes, Inc. Used by permission; all rights reserved.*

Accessible patio areas are available for meetings or for relaxation. *Courtesy of: Roofscapes, Inc. Used by permission; all rights reserved.*

Industrial/Commercial: Intensive

601 Congress Street, Seaport District

Location: Boston, Massachusetts
Project Type: New construction
Green Roof Size: 11,000 square feet
Completion Date: 2004
Year of Award: 2006

Client/Developer: Manulife Financial
Project Architect: Skidmore, Owings & Merrill, LLP.
Landscape Architect: Sasaki Associates, Inc. *(Winner)*
Landscape Contractor: ValleyCrest Landscape Development

Green Roof System Details
Waterproofing: American Hydrotech, Inc. 215 mils hot rubberized asphalt in a protected membrane roof assembly.
Drainage: American Hydrotech drainage board.
Irrigation: Drip irrigation.
Growing Medium: 6"-12" depth composed of 55 per cent rotary kiln expanded lightweight aggregate (expanded shale), graded sand, and treated compost derived from cranberry waste.
Vegetation: Drought tolerant ornamental grasses and sedums (*See Appendix B: Plant Lists*)
Design Objectives: Aesthetic improvement; new amenity spaces; storm water management: quantity; low maintenance.

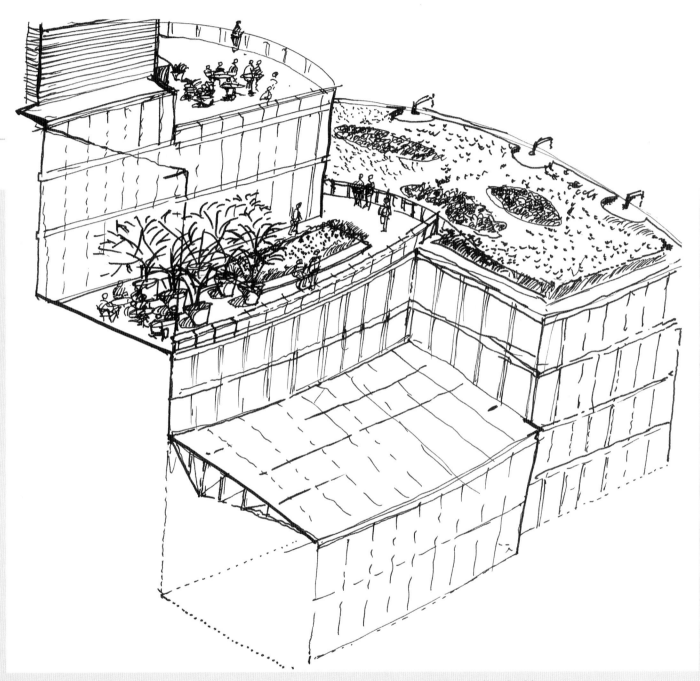

Separate amenity spaces allow building occupants to enjoy the green roof and the ocean view beyond.
Courtesy of: Valleycrest Companies

As an organization, Manulife Financial is committed to enacting the principles of sustainable building. When they undertook the task of building new offices in Boston's harbor, there was no doubt that the structure should meet LEED® standards. *(See Appendix D: Green Roofs and Leadership in Energy Efficient Design)* The twelfth floor green roof was incorporated into the project as a key design element used to achieve this goal.

Brought into the project by Skidmore, Owings & Merrill, LLP, the landscape architects at Sasaki Associates, Inc. took the opportunity to create a meadow that served as a foreground to the magnificent ocean view. A glassed-in terrace provides seating space that overlooks the surrounding city and its harbor. A glass railing separates the useable paved area from the planted one without marring the visual effect of the plants' forms, textures, and seasonal colors.

The green roof was included in the project late in the design phase. As a result, its design had to accommodate the limitations of an already existing roof design. For example, parapet walls were not high enough to meet safety requirements, causing the accessible portion of the roof to be restricted to a smaller paved area. However, this proved to be an aesthetic boon since it allowed the green roof to blend seamlessly into the ocean background.

The low parapets also increased the roof's exposure to the elements and made for very harsh environmental conditions. The presence of drying winds necessitated the selection of drought tolerant plant materials that could cover and shade the ground plane to prevent extensive water loss through evaporation. The planting scheme is composed primarily of natural grass masses that vary in height from one to two feet. These masses are punctuated by drifts of taller, ornamental Miscanthus grass. In protected areas near the building's interior walls, beds of low-growing sedums provide a variety of colors and floral texture.

To avoid excessive wind erosion, plugs were planted densely and covered with a jute blanket. Despite these preventative measures, some erosion did occur in the first winter, but since the plant establishment period has ended wind erosion has not been a problem. Similarly, the weight of the growing medium was of great importance to avoid the occurrence of wind uplift. The vegetation selection minimized the need for maintenance, which consists of pruning the ornamental grasses to a six-inch height every spring and replenishing depleted nutrient levels in the growing medium via the irrigation system once or twice a year.

This green roof was designed for functional purposes, yet it is also serves an equally important, aesthetic purpose: reconnecting the structure and its occupants with the surrounding environment. The aesthetic and psychological benefits of this space are as great as the ocean it overlooks.

"Green roofs are not just a technical solution. Their design need not be limited by their function; they can have imaginative content which visually connects them with their surroundings."
— Alistair McIntosh, Landscape Architect, Sasaki Associates, Inc.

Industrial/Commercial Projects and Award Winners   140

Low parapet walls allow the vegetation to meet the horizon seamlessly.
Courtesy of: Valleycrest Companies

Tall grasses create textural variation even in the winter months.
Courtesy of: Valleycrest Companies

Industrial/Commercial: Intensive

ABN AMRO Plaza

Location: Chicago, Illinois
Project Type: New construction
Green Roof Size: 74,700 square feet
Completion Date: 2003
Year of Award: 2007

Client/Developer: Gerald Hines Interests and LaSalle Street Capital
Architect: De Stephano and Partners, Ltd.
Landscape Architect: Daniel Weinbach & Partners
Roofing Consultant: Barrett Company *(Winner)*
Landscape Contractor: Christy Webber Landscape

Green Roof System Details
   Waterproofing: Barrett Company 215 mil rubberized asphalt membrane in a protected membrane roof assembly.
   Irrigation: Pop-up sprinklers and drip irrigation.
   Growing Medium: 18-36" of Midwest Trading brand general purpose amended top soil blend.
   Vegetation: A variety of trees, shrubs, and groundcovers. (*See Appendix B: Plant Lists*)
   Design Objectives: Aesthetic improvement; new amenity spaces (recreation); storm water management: quantity; reduction of the urban heat island effect.

Industrial/Commercial Projects and Award Winners

Mature plant life allowed the client to enjoy an instant green effect. *Courtesy of: Barrett Company*

The green roof and terrace areas on the 6th floor podium of the ABN AMRO Plaza at 540 West Madison Avenue in Chicago, were built primarily as amenity spaces for the enjoyment of employees. This is the first major green roof installation by the noted developer, Gerald Hines Interests of Houston, Texas, on one of their high-end commercial properties. Though the developer had been toying with the idea of green roofs for some time, it was not until they were introduced to the City of Chicago's strong commitment to this technology that all doubts were dispelled.

One of the causes for hesitation was related to waterproofing: would the additional materials increase its likelihood of leaking? A waterproofing membrane, when properly designed and installed, will in fact last longer in a green roof than in a traditional roofing application. The green roof will preserve the integrity of the roof membrane by eliminating damage from the sun's ultraviolet light, eliminating thermal shock, and, if properly protected during implementation, all mechanical damage. Most waterproofing membranes, in this case monolithic rubberized asphalt, can withstand constant water immersed conditions and hydrostatic pressures while providing long term physical and chemical stability in an environment subjected to water, fertilizers, root growth, and chlorides.

Having worked on many other roofs like this one, the implementation process was fairly simple for an experienced green roof installation company like Barrett Company. There were a few tricky details around some counterflashings, the formed metal secured to a wall, curb or roof top unit to cover and protect the upper edge of a base membrane flashing and its associated fasteners, and planter drains. Through the dedication of and clear communication with Rebecca Callcott of De Stephano and Partners, Ltd., these design details were sorted out.

Designed for maximum accessibility, one half of the roof is green and the other is made of pavers and planters of various heights. The planted area includes a diverse array of vegetation that is complimented by an open lawn planted with sod. Some of the trees were fairly mature when they were planted, giving the green roof an immediate true garden feel. Maintenance of these planted areas is fairly intensive and consists of a spring clean up, followed by weekly care from April to September. It is completed by a clean up period in the fall.

This project is an important addition to Chicago's green roof success story. The 6th floor podium on the ABN AMRO plaza contributes to Mayor Daley's objective of making Chicago a cleaner, healthier, more energy-efficient environment for its citizens. It demonstrates the potential of green roofs to provide a pleasing, tree-lined refuge in the center of the bustling city.

"It is highly indicative of current building trends that green roofs are being implemented not only by institutions and public facilities but also by one of the most astute developers of private office space."

— Tim Barrett, Roofing Consultant, Barrett Company

The roof is a tree-lined refuge in the center of a bustling city.
Courtesy of: Barrett Company

Industrial/Commercial: Extensive

Calamos Investment

Location: Naperville, Illinois
Project Type: Retrofit
Green Roof Size: 3,088 square feet
Completion Date: Summer 2006
Year of Award: 2007

Client/Developer: Calamos Investments
Architect: Lohan Anderson
Landscape Contractor: Intrinsic Landscaping, Inc. *(Winner)*

Green Roof System Details
  Waterproofing: American Hydrotech, Inc. hot applied rubberized asphalt in a protected membrane assembly.
  Drainage: American Hydrotech, Inc. drainage board.
  Growing Medium: 4" depth of Hydrotech Lite Top.
  Vegetation: 12 different sedum varieties. (*See Appendix B: Plant Lists*)
  Design Objectives: Aesthetic improvement; new amenity spaces (recreation); low maintenance.

Industrial/Commercial Projects and Award Winners

The rooftop patio provides a place where people can retreat to relax and reduce stress levels during the working day. *Courtesy of: Intrinsic Landscaping, Inc.*

After having built a new multimillion-dollar building, John P. Calamos, Sr., President of Calamos Investments, a diversified global investment management company, looked at his 10th floor patio roof and was confronted by a stark expanse of gravel. He promptly called the building's architect, Dirk Lohan of Lohan Anderson, who then contacted Intrinsic Landscaping Inc., to explore the available options. They felt a green roof would be an excellent choice and decided to go with a vegetative mat system with over four inches of growing medium due to its "instant green" effect.

Though the installation process was quite straightforward, the green roof's ability to satisfy the design goals hinged on the active dialogue between all parties involved. The roof had been waterproofed with an American Hydrotech, Inc. system, which is often used in green roof assemblies. This happy coincidence allowed the roof to retain its warranty and sidestep a potential obstacle to its greening. The membrane was carefully inspected to ensure that it was leak free before the drainage layer, root barrier, insulation, aeration layer, moisture retention mat, flora drain, system filter, and growing medium were laid in place. The green roof was finally completed when the vegetative mats, which support twelve different species of sedum, were rolled out.

The green roof offers an attractive view from within the work place; it can be viewed from an accessible patio area or from conference rooms and executive offices. The presence of plants and green space has proven to be helpful for improving employee productivity in the work environment. The rooftop patio provides a place where people can retreat to relax and reduce stress levels during the working day.

The green roof provides added insulation to the building, reducing energy costs and ensuring greater roof longevity by protecting the membrane from the elements. It will also minimize storm water runoff and improve air and water quality, while providing habitat for birds and insects. Lower surface temperatures on the green roof will help to mitigate the local effects of the urban heat island, reducing contributions to global warming.

The Calamos Investments project is an excellent example of a simple and elegant green roof retrofit, providing a visible and accessible replacement for an otherwise unattractive and undistinguished rock ballast roof. It also highlights the market potential of extensive green roofs in the commercial and industrial sectors. By creating a green roof that is visible and accessible for a company on the forefront of real estate and development, Calamos Investments is well positioned to enjoy its many benefits and incorporate the idea into future projects!

"This application really shows the value of vegetative mats. The green roof looked good the minute we walked off the roof because the plant coverage was dense. Density like that takes years to achieve in roofs that are planted in place."

— Kurt Horvath, Landscape Contractor,
Intrinsic Landscaping Inc.

The roof replaces the gravel expanse that was once outside the building's boardrooms.
Courtesy of: Intrinsic Landscaping, Inc.

The use of vegetative mats allowed the design team to immediately satisfy the client's aesthetic demands.
Courtesy of: Intrinsic Landscaping, Inc.

Civic Awards: Championing the Cause

Mayor Richard M. Daley shows French Minister of Foreign Affairs Philippe Douste Blazy Chicago's Green Roof Demonstration Project. *Courtesy of: The City of Chicago.*

Champions drive the green roof policy process forward, which is critical to the widespread adoption of the technology. Public policy support helps to overcome the higher initial capital costs of green roofs and recognizes their tangible public infrastructure benefits. Typically, there are one or more champions in the community from the government and/or non-profit sector that work to initiate the policy development process. They may be politicians, bureaucrats, academics or representatives from non-profits organization. Each year Green Roofs for Healthy Cities acknowledges the outstanding contribution of a civic leader to green roof development through our *Civic Award of Excellence Program*.

Mayor Richard M. Daley
City of Chicago, Illinois

Year of Award: 2003

Chicago is a rare case where interest in green roof policies began at the very top. After seeing the positive impacts of green roofs on a 1998 tour in Europe, Mayor Richard M. Daley requested that the Chicago Department of the Environment examine the feasibility of putting a green roof on top of City Hall. This resulted in the implementation of the City Hall demonstration project: a rooftop garden with over 20,000 plants comprised of 156 species in early 2000.

Test sites at the Chicago Center for Green Technology were subsequently constructed and used to quantify thermal storm water runoff. Additional plots were installed in a public place, the Garfield Park Conservatory, to display horticultural possibilities and counteract the notion that extensive green roof technology cannot be aesthetically pleasing. Both demonstration projects continue to be in effect and their benefits can still be accrued by the public.

With the assistance of the Chicago Urban Land Institute, a local real-estate and land use not for profit organization, the Chicago Department of Planning and Development (CDPD) held a series of seminars and focus groups in March 2003 to educate and gauge public, political, and industry opinion about green roofs. Most participants were hesitant to embrace the idea because of the scarcity of local data, but were interested in some kind of incentive program. In 2003, the City of Chicago co-hosted the *First Annual International Greening Rooftops for Sustainable Communities Conference, Awards and Trade Show*, the green roof industry's first major conference.

Since that time Chicago has developed the most comprehensive suite of green roof strategies in North America. These include regulatory measures, financial incentives, and awareness raising strategies:

- The Chicago Standard and Building Green/Green Roof Matrix
- Green Permit Program
- Green Urban Design
- Green Roof Grants Programs
- Green Roof Improvement Fund
- Storm Water Management and Water Agenda
- Green Roof Web Resources
- Green Roof Request For Information (RFI)
- Guide to Roof Top Gardening

> "*Although my first thoughts of vegetated rooftops occurred in 1990, it was another 6 years before I installed an ecoroof on my garage. And it was my garage ecoroof that provided the proof that vegetated roof systems work! The projects in this book are like a dream coming true, a dream of urban environments that are healthier and more livable, not just for people but other life forms so essential to earth. The contributive efforts by all those involved in this fledgling green roof industry are indeed very important to our future.*"
>
> — Tom Liptan, City of Portland

Tom Liptan, Bureau of Environmental Services
City of Portland, Oregon

Year of Award: 2004

Tom Liptan was initially exposed to the ecological functions of green roofs in 1990 during a presentation in Germany. Four years later, he bought a new dish soap whose label boasted about the greenness of its production facilities, including a green roof. Intrigued by the concept, he called the soap's sustainability-minded Belgian manufacturer Ecover, and soon came to understand the beneficial relationship between green roofs and storm water management. Two years later a green roof was installed on his garage.

Tom quickly began to apply his enthusiasm for green roofs to his work on storm water management within Portland's Bureau of Environmental Services. Discussion in the Bureau, extensive research, and monitoring runoff quality and quantity from his garage eventually led to the financing of a demonstration project atop the Hamilton Apartment Building in 1999. This project was used to monitor storm water flows and precipitation, as well as to determine plant suitability to the area's hot and dry summers.

Since public support was first given to the project by the Bureau's commissioner, Portland's green roof industry has developed quite rapidly. In 2001, a Floor-Area Ratio (FAR) incentive program was put in place to encourage implementation. The program has been quite popular and is currently undergoing revision. Furthermore, city specific green roof design guidelines are under development.

This cross-departmental effort has resulted in over 400 built or planned public and private green roofs in the metropolitan area, equaling over four million square feet of green space.

Policy Development Timeline

- **1999-** City Hall Commissioning/Design
- **2000-** Implementation of City Hall Green Roof
- **2002-** Urban Heat Island Grant Program
- **2003-** Co-Host First Green Roof Conference
- **2003-** Research on Green Roof Systems
- **2004-** Green Building/Green Roof Policy Matrix
- **2004-** Millennium Park Opens
- **2005-** DCAP Green Permit Program
- **2005-** Green Roof Residential Grant Program
- **2005/2006-** Green Roof Web Site
- **2006-** Tax Incentive Financing in LOOP
- **2006-** Renewed Residential Grant Program

> "*We do this not because it's fashionable, but because it makes sense. It improves public health; it beautifies the city; it enhances the quality of life; it saves money; and it leaves a legacy for future generations.*"
>
> — Mayor Richard Daley, City of Chicago

> "Green roofing is the only technology that can provide such a wide array of environmental benefits. They are an important tool that can be used to reshape the urban landscape for a greener future."
>
> — Karen Moyer, City of Waterloo

Green roofs, locally called "eco-roofs" still face barriers to wide spread dissemination: the first being that the FAR is currently only applicable to the Central District and therefore ignores a large part of the market; the second is that many manufacturer's pre-designed systems are not able to withstand the summer heat without irrigation. Further research needs to be conducted into what green roof components (e.g. irrigation and vegetation) are best suited to the local climate.

Until they are institutionalized, Tom believes that eco-roofs still require the support of a championing Bureau. Though green roofs are not yet commonplace, it is clear that they are in Portland to stay.

Policy Development Timeline

1999- Hamilton Apartment Building demonstration project

2000- Monitoring roof for storm water runof quantity and quality

2001- Floor-Area Ratio, a density bonusing incentive program, introduced in the Central District

2001- Incorporated into the Central City Fundamental Design Guidelines

Spring 2004- City host *Second Annual International Greening Rooftops for Sustainable Communities Conference, Awards and Trade Show.*

Summer 2005- City's Green Building Policy requires that green roofs be considered for all projects and implemented when feasible.

Late 2006- Introduction of the Clean River Awards program allows properties with onsite storm water management techniques to qualify for up to a 100 per cent discount on storm water utility fees.

Karen Moyer, Environment Special Projects Manager, Capital Projects and Services City of Waterloo, Ontario

Year of Award: 2005

During the development of Waterloo's Environmental Strategic Plan green roofs were perpetually being brought to Karen Moyer's attention by professionals spanning several disciplines as a means of addressing the issues of air quality, green space, community building capacity, and the rehabilitation of Laurel Creek. In 2004, Karen commissioned a Green Roof Feasibility Study that was subsequently funded with 25,000 dollars from the Federation of Canadian Municipalities.

The objective of the study was to determine the specific benefits of green roofs for Waterloo while identifying areas that would most benefit from implementation of the technology. One of the Feasibility Study's goals was to build a green roof demonstration site on a city building. The feasibility study broke new ground using Geographic Inforamtion Systems (GIS) to map out areas that would reap the maximum benefit from various green roof attributes. In December of 2004 the finished report was submitted to council and was approved in April 2005. A 1,500 square foot demonstration green roof site was planned for Waterloo's City Hall and over thirteen projects were installed in the following two years.

Karen Moyer became acquainted with Green Roofs for Healthy Cities during the process of completing the Feasibility Study. On February 6th, 2003, Karen spoke about the Waterloo experience at Green Roofs for Healthy Cities' Regina Symposium, and again in April 2003 during the Waterloo Symposium. Karen has been actively involved in

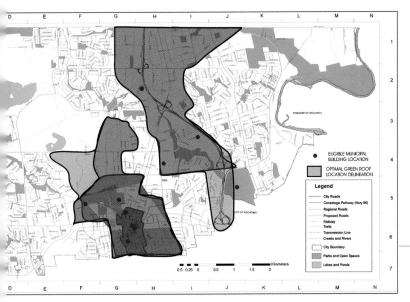

The City of Waterloo mapped the energy efficiency, storm water and air quality benefits of green roofs across the city.
Courtesy of: The City of Waterloo

"The benefits to green roofs are numerous, especially related to the urban heat island effect and energy use. Hopefully a time will come when, through technologies like green roofs, we can avoid unnecessary energy creation."
— Deputy Mayor Joe Pantalone,
City of Toronto

every stage of green roof infrastructure development in her community from community engagement and action plan development to technical research and policy development. Karen received this award for her professionalism passion and perseverance.

Policy Development Timeline

April 2003- Green Roofs for Healthy Cities Local Market Development Symposium

2004- GreenRoof Feasibility Study and City Wide Implementation Plan commissioned and completed

April 2005- Recommendations from study approved by City Council

September 2005- Demonstration project completed on The City of Waterloo City Centre

Councillor Joe Pantalone, Deputy Mayor City of Toronto, Ontario

Year of Award: 2006

Deputy Mayor Joe Pantalone is a political veteran with a compelling track record in serving the City of Toronto. He is well known for his dedication to a beautiful and safe city, efforts in greening the city, and service to constituents. As Chair of the Board of Governors for Exhibition Place, Joe has worked tirelessly to make the site an environmental, heritage, and festival showcase. Besides pioneering the position of the City's Tree Advocate, which plants tens of thousands of new trees in Toronto yearly, Joe, as Chair of the Roundtable on the Environment, is leading Toronto's efforts to become North America's leader in the environmental field.

His enthusiasm for ground level greening was translated to the roof when he saw a presentation about green roof technology. At that presentation, he realized that a huge portion of the earth's surface is ignored, and badly treated at that. This galvanized him into action and propelled green roofs onto the agenda of Toronto's *Environmental Plan.*

As Chair of the Roundtable on the Environment, Joe enthusiastically drove the development of green roof policies and research at City Hall. In 2004, with a grant from the Federation of Canadian

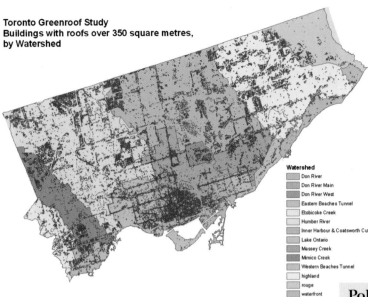

Toronto Greenroof Study
Buildings with roofs over 350 square metres, by Watershed

The *Report on the Environmental Benefits and Costs of Green Roof Technology for the City of Toronto* utilized a GIS system to map the impact of greening the 100 percent of the available city-wide roof area (represented by black dots), which included flat roofs on buildings with more than 350 square meters of roof area, and assuming at least 75% of the roof area would be greened. The total available green roof area citywide was determined to be 5,000 hectares (50 million square meters). *Courtesy of: Dr. Hitesh Doshi, Ryerson University*

Municipalities *Green Municipal Funds* and Earth and Environmental Technologies, the City was able to commission a study of municipal level cost-benefit analysis. The 2005 *Report on the Environmental Benefits and Cost of Green Roof Technology for the City of Toronto* utilized a model that assumed at least 75 per cent green roof coverage on existing flat roofs to help quantify the benefits to the City.

A subsequent report, *Making Green Roofs Happen,* incorporated feedback from stakeholder forums and outlined the Roundtable on the Environment's preliminary recommendations to City Council. This issue was the first "good news story" that had filled the house at a public meeting; over 200 people were present, 31 of which took to the stand and lauded the City's interest in the technology. The report contains over 20 recommendations ranging from procurement to research to various green roof incentive programs.

Within several months, the City launched the *Green Roof Incentive Pilot Program*, providing a financial incentive of ten dollars per square meter with an upper limit of 20,000 dollars per project. The next year's incentive was increased to 50 dollars a square meter. The public incentive was complement by an internal green procurement by government facilities, requiring that green roofs be installed on new City owned buildings and on retrofit projects where feasible. This process also laid the groundwork for the development of additional policies.

Councillor Joe Pantalone, with the support of the Roundtable on the Environment and Mayor David Miller, was awarded the *Green Roof Civic Award of Excellence* because his strong belief in the benefits of green roofs and his tireless efforts to create a Toronto with cleaner air and water, new amenity space, and beautiful green vistas.

Policy Development Timeline

November 2000- City Hall green roof demonstration project completed with GRHC, Federal and Local government funding; research initiated.

Spring 2003- Green Roof research on Urban Heat Island benefits, City Hall and Eastview completed by Environment Canada, NRC's Institute of Research in Construction and City of Toronto.

Summer 2005- Study period of the multiple benefits of widespread green roof installation on existing buildings with City of Toronto and Ryerson Unversity.

September 2005- Focus group workshops held by City of Toronto.

October 31st 2005- Publication of Ryerson University *Report on the Environmental Benefits and Costs of Green Roof Technology for the City of Toronto* and submission of a draft report on green roof strategies.

November 23rd 2005- Meeting Environmental Roundtable Town Hall meeting to discuss Making Green Roofs Happen: A Discussion Paper Presented to Toronto's Roundtable on the Environment.

February 2005- Roundtable Recommendations for Toronto's Green Roof Strategy adopted by the Policy and Finance Committee.

April 2006- Implementation strategies brought before Toronto City Council.

Spring 2006- Green Roof Pilot Program launched. It is comprised of a $10 per square meter grant.

July 2006- City Council passed a series of recommendations towards the creation of Toronto Green Development Standard, including green roofs and walls.

Summer 2007- Grant program increased to $50 per square meter.

" I first became excited about green roofs when I looked out of the window of a hotel in our downtown and saw all of the barren rooftops in our city. It made me realize that there was a really big opportunity here, and with the right incentives, many developers would choose to make rooftops a valuable asset for our community, to help clean the air, better manage stormwater and improve the aesthetics in our city."

— Council Member Lisa Goodman, City of Minneapolis

Council Member Lisa Goodman
City of Minneapolis, Minnesota

Year of Award: 2007

Changing longstanding practices and public policies does not happen easily, and it rarely happens in government without either a crisis or a committed champion. In Minneapolis, that champion has been Council Member Lisa Goodman who has led the City's efforts to promote green roofs on a number of fronts. Her efforts have made Minneapolis a leader in the promotion and implementation of green roofs both within the city and throughout the region.

In 2004, Council Member Goodman was introduced to green roofs and their potential to improve the City. The next summer, she used funds from her office budget to hire a graduate student at the University of Minnesota College of Landscape Architecture to work on green roof issues. At the time, the City was working to establish a new storm water utility that would impose fees on property owners based on their impact on the municipal storm water system.

The storm water utility model was an attempt to align the costs of maintaining the public infrastructure for storm water with the amount of storm water coming from each piece of property. Council Member Goodman saw an opportunity for the new model to push property owners to incorporate green roofs as part of their developments and to reduce the new storm water utility fees for properties that included green roofs. She pushed for the project to include a 100 per cent fee abatement, the most aggressive fee-based incentive of any major city in the country, and without her role as champion, it would not have happened. As adopted policy, the City will abate the storm water fee if a development project has a best management practice with both a quality and quantity reduction in storm water runoff.

Since then, Council Member Goodman has advocated to include green roofs as an alternative compliance to the City's at-grade greening requirements and supported efforts by the City Planning Division to draft performance standards for green roofs that clarify how they could be evaluated as part of site plan reviews and other aspects of the City's land use process. She also supported a feasibility study on constructing a three-acre-plus green roof on the Target Center, the City-owned arena in downtown and led efforts to construct an advanced demonstration green roof at City Hall.

Currently, she is the Chairperson of the City Council's Community Development Committee and the Minneapolis Community Development Agency. During her tenure, she has led the successful effort to merge the City's Planning Department and Development Agency to form the Community Planning and Economic Development Department (CPED).

Policy Development Timeline
Summer 2005- Dedicated staff person hired to examine green roof issues
2005- Introduction of new storm water utility model, including green roofs as a Best Management Practice
May 2005- Community Planning & Economic Development, Planning Division include green roofs in Alternative Compliance Clause of the *Site Plan Review*.
Spring 2007- Demonstration project at City Hall complete
May 2007- City hosts 5th *Annual Greening Rooftops for Sustainable Communities Conference, Awards, and Trade Show*

Research Award: Developing Performance Knowledge

Opposite: Center for Green Roof Research at Penn State University. *Courtesy of: Dr. Robert Berghage, Center for Green Roof Research, Penn State*

The Research Award of Excellence was established to recognize significant contributions to our understanding of green roof performance in North America. Ongoing research at the system, building, site, and community scales is required to fully appreciate the full extent of green roof performance benefits. Biophysical and economic green roof research supports product development, the integration of green roofs with other building systems, the development of analytical tools, the establishment of regionally tailored systems and the development and implementation of public policy support. This award, established in 2006, acknowledges outstanding professionals in this rapidly growing field.

Dr. David Beattie, Associate Professor of Ornamental Horticulture Pennsylvania State University, Founder of the Penn State Center for Green Roof Research

Year of Award: 2006

Dr. David Beattie is the Founder and Director of the Penn State Center for Green Roof Research at Pennsylvania State University and an Associate Professor of Ornamental Horticulture. He received his B.S. in Horticulture at Rhode Island University, an M.S. in Horticulture at the University of Vermont, and completed his Ph.D. at Michigan State University. During his career, Dr. Beattie's work has been extensively published in several academic journals, research papers, and books. His research interests include green roofs, perennial plant growth, their development and use as flowering pot plants, and root growth and control in pots (the effect of copper and growth retardants on root architecture).

Through his multiple applied and basic research projects, Dr. Beattie has played a significant role in shaping and informing the green roof industry. In 2000, he founded the highly regarded Penn State Center for Green Roof Research with a mission to promote green roof research, education, and technology transfer in the northeastern United States. He assembled a team of scientists from Horticultural and Agricultural Engineering backgrounds who then collaborated on numerous research projects. Under his direction, the Center has produced some of the most complete and comprehensive green roof research in North America.

The Center has been involved in a variety of research projects demonstrating the potential for green roofs as a storm water best management practice in North America. The Center has collected performance data from their green roof structures over the last six years, which indicates that extensive green roofs will retain approximately 40-50 per cent of the annual precipitation and can improve the quality of water runoff.

The facility has six structures, which are insulated, heated, and air-conditioned. Three have green roofs, and three are fitted with traditional roofs. A total of thirty-nine sensors are installed in the green roof structures and fourteen in the control group. Data is collected every half hour. The structures are fitted with household watt-hour meters, which read daily, and thermisters that measure heat flux in the floors, walls, and ceilings. The knowledge gathered from these readings will be used to create a computer model to enhance

current green roof technologies and energy saving measures.

The Center has also led the way in developing media and membrane testing protocols based on pre-existing German standards. As a result Penn State is now offering commercial green roof media testing through the Penn State Agricultural Services Laboratory and has begun testing commercial roofing membranes for their resistance to root penetration.

Other research projects being conducted at the Center for Green Roof Research include: water quality studies analyzing storm water runoff for pH, turbidity, and nitrates; evapotranspiration model development with preliminary estimates that some plants exclusive of growing media can absorb up to 25 per cent of precipitation; media analysis aimed at replicating FLL standards; and, studies on plant growth and speed.

As part of his desire to share his green roof knowledge with others, David continually communicates his research findings through research papers, brochures, books, presentations at various conferences and seminars, teaching green roof courses, and a website. He has also assisted, advised, or co-advised graduate students working on green roof projects.

David Beattie's green roof research and the ongoing efforts of his successors continue to contribute to the depth of knowledge and general awareness of green roof technologies in North America. The broad range of his research and commitment to sharing his knowledge with others has inspired his students and peers who will carry his work forward for generations to come.

Bibliography

Bass, Brad; Krayenhoff, E. Scott; Martelli, Alberto; Stull, Roland B.; and Auld, Heather. "The Impact of Green Roofs on Toronto's Urban Heat Island." *First Annual Greening Rooftops for Sustainable Communities Conference*. Green Roofs for Healthy Cities: Chicago IL, May 2003.

Casey Trees/Limno Tech. Regreening Washington, D.C.: A Green Roof Vision Based on Quantifying Storm Water and Air Quality Benefits. Washington, D.C, 2005. http://www.greenroofs.org/resources/greenroofvisionfordc.pdf

City of Waterloo. Green Roofs Feasibility Study: A City Wide Implementation Plan. Final Report December 2004. http://www.city.waterloo.on.ca/DesktopDefault.aspx?tabid=2014

Doshi, Hitesh; Baniting, Doug; Li, James, Missios, Paul; Au, Angela; Currie, Beth Anne; Verrati, Michael. (November 2005). Report on the Environmental Benefits and Costs of Green Roof Technology for the City of Toronto. Policy and Research, City Planning Division, City of Toronto. http://www.toronto.ca/greenroofs/findings.htm

Green Roofs for Healthy Cities. Final Report: Green Roof Industry Survey 2004 & 2005. Green Roofs for Healthy Cities: Toronto, ON, 2006

Kellert, Stephen. *Building for Life: Designing and Understanding the Human-Nature Connection*. Island Press: Washington, D.C., 2005.

Koehler, M., M. Schmidt, M. Laar, U. Wachsmann, and S. Krauter. "Photovoltaik-Panels on Greened Roofs." Krauter (ed). *World Climate & Energy Event*. Rio, Brazil: Dec. 1-5, 2003. p. 151-8.

Leonard, Timothy and James Leonard. (2005) "The Green Roof and Energy Performance – Roof top Data Analyzed." *Third Annual Greening Rooftops for Sustainable Communities Conference*. Green Roofs for Healthy Cities: Washington, D.C., May 2005.

Liu, Anthony. Ontario's Electricity Demand Response to Climate Change. Unpublished by Environment Canada: Toronto, 2006.

Liu, Karen and Baskaran, Brad. "Thermal Performance of Green Roofs Through Field Evaluation." *First Annual Greening Rooftops for Sustainable Communities Conference*. Green Roofs for Healthy Cities: Chicago, IL, May 2003.

McDonough, William and Braungart, Michael. *Cradle to Cradle: Remaking the Way We Make Things*. North Point Press: New York, NY, 2002.

McLinden, Steve. "Eco-Friendly Apartments Get the Green Light." *National Real Estate Investor*. (January 1 2004) Accessed on-line, 2006, www.nreionline.com/property/multifamily/real_estate_ecofriendly_apartments_green/.

Policy and Research, City Planning Division, City of Toronto. "Section 5: Conclusions." *Making Green Roofs Happen: A Discussion Paper Presented to Toronto's Roundtable on the Environment*. Toronto, Ontario, 2005. http://www.toronto.ca/greenroofs/policy.htm

Rosenzweig, C., & Solecki, W., et al., *Mitigating New York City's Heat Island with Urban Forestry, Living Roofs, and Light Surfaces*. Columbia University and Hunter College: New York, 2005. http://ams.confex.com/ams/pdfpapers/103341.pdf

Rosenzeig, Michael L. *Win-Win Ecology: How the Earth Species can Survive in the Midst of Human Enterprise*. Oxford University Press: New York, 2003.

Wells, Malcolm. *Recovering America: A More Gentle Way to Build*. Brewster MA, 1999.

Appendix A:
About Green Roofs for Healthy Cities

www.greenroofs.org

Green Roofs for Healthy Cities – North America, Inc. was founded in 1999 as a small network of public and private organizations and is now a rapidly growing 501(c)(6), not-for-profit industry association. Our mission is to increase the awareness of the economic, social and environmental benefits of green roofs and green walls, and other forms of living architecture through education, advocacy, professional development, and celebrations of excellence.

We work tirelessly with our members to raise awareness of green roof technology through several channels: training and accreditation program, annual conference, awards of excellence and through our quarterly magazine, The Living Architecture Monitor.

We deliver numerous courses across North America covering all aspects of green roof projects, ranging from basic design issues to implementation to policy to maintenance. Our courses qualify for continuing education credits with many organizations, including the American Institute of Architects, American Society of Landscape Architects, and RCI, Inc. Please check our website (www.greenroofs.org) for course schedules.

Design 101 Introductory Course (2nd Edition)

This Introductory Course teaches you about the components and benefits of green roofs, how to design a project for maximum client benefits, how to cost estimate for different types of projects, how to incorporate important design principles and design for maximum LEED® credits. This course includes the most up-to-date green roof research and design practices as well as award winning case studies from this rapidly expanding industry.

Green Roof Infrastructure: Design and Installation 201

Design 201 covers in detail the many different steps that must be understood and incorporated into any successful green roof design and implementation, including actors' roles and responsibilities, construction contracts and their administration, and how to avoid common installation errors. This course is suited to those who have already taken Design 101 or who already have a good general understanding of green roof design.

Green Roof Infrastructure: Waterproofing and Drainage 301

This course provides an overview of waterproofing and drainage construction and maintenance for green roof assemblies. It lays out the technical vocabulary and the various benefits of different waterproofing materials and presents detailed design solutions and implementation best management practices for waterproofing and drainage in green roofs.

Green Roof Infrastructure: Plants and Growing Medium 401

This course provides an overview of plants and growing medium selection, installation, and maintenance for green roof assemblies. It lays out the technical vocabulary and various design approaches employed to optimize the use of these materials and presents implementation best management practices for plants and growing medium in green roofs.

The accreditation program focuses on multidisciplinary knowledge and skills training and development. *Accredited Green Roof Professionals* (e.g. architects, landscape architects, engineers, contractors, manufacturers, and roofing consultants) will have knowledge of the best practices associated with the design, installation and maintenance of green roofs and berequired to satisfy an ongoing continuing education component. Accreditation will be available in the marketplace in 2009.

The *Annual International Greening Rooftops for Sustainable Communities Conference, Awards, and Trade Show* was developed as a forum for information exchange in this rapidly growing industry. Held in a different city every year to develop different markets, conference participants are presented with the most up-to-date research, policy, and design trends in North American green roofing.

On a smaller scale, we work in partnership with various municipalities to create an action plan for

Green Walls 101: Introduction to Systems and Design

This full day course covers the many benefits and design and implementation strategies for vertical greening technologies which allow plants to grow on the exterior or interior of buildings.

Ecological Green Roof Design Workshop

This half day workshop acquaints participants with ecological design principles and discusses their application in the context of green roofing so as to maximize this much sought after benefit.

Green Roof Policy Development Workshop

This half-day workshop examines what kind of incentives currently exist in other jurisdictions, how the incentives are tied to performance standards and how policy makers can assess what types of incentive programs will work best in their areas. It also provides a framework for planners and governmental agencies to gain an understanding of the multi-faceted impacts of green roofs and areas of consideration for implementing green roof policy.

Our *Accredited Green Roof Professional Designation Program*, currently under development through a multi-stakeholder process, is the first of its kind. It will:

Allow professionals to differentiate themselves in the marketplace.
Establish a high level of professionalism.
Increase customer confidence in green roof technology.
Protect public health, safety, and welfare.
Result in better green roof design and installation practices.

policy development tailored to the local political context through the organization of full-day *Local Market Development Symposia*. These events provide an opportunity for participants to learn from local and national experts about the general design and implementation of green roofs.

The *Living Architecture Monitor* is new – but only in name. The *Green Roof Infrastructure Monitor* was first launched by Green Roofs for Healthy Cities as a four-page newsletter in 1999 and rapidly blossomed into a full-blown magazine published twice per year. However, as the green roof industry grows in new and exciting ways, we realized the need to evolve our magazine to better meet the needs of an increasingly sophisticated and complex marketplace. The new name – *Living Architecture Monitor* – reflects our enhanced commitment to educate, inspire, and celebrate this emerging restorative, sustainable vision of architecture and landscape design that extends beyond green roofs.

As the nonprofit association for the green roof industry in North America, we have worked hard with our Corporate and Individual Members to raise the bar in terms of professionalism and excellence though tools, like the *GreenSave Calculator* and *Green Roofs Tree of Knowledge*, learning opportunities and inspiration including the facilitation of research and access to green roof research. Our membership is comprised of individuals and companies from many disciplines, professions, and sectors spanning multiple geographic regions; it is diversity that makes this industry so interesting, exciting, and unique.

The *GreenSave Calculator* is a web-enabled Life Cycle Cost Benefit Calculator that focuses on long time frames, real monetary costs and savings, and financial returns attributed to employing conventional and green roofs. It allows green roof designers the opportunity of including a wide range of quantifiable benefits into the life cycle model by utilizing their own, or some of the provided, reference values.

The *Green Roofs Tree of Knowledge* is a full-featured online research and policy database related to green roof and green wall infrastructure. There is a considerable amount of work being done on the many socioeconomic and biophysical benefits that green roofs and other forms of living architecture provide. This database is composed of detailed summaries of research and policy papers written by expert contributors from around the world.

Through these tools, we continue to ensure green roof information is widely available and easily accessible to inspire individuals to do their very best work. In this way, we also promote and celebrate excellence in the green roof industry in North America.

The *Green Roof Awards of Excellence* was launched in 2003 to recognize green roof projects that demonstrate extraordinary leadership, recognizing and celebrating the valuable contributions of green roof design professionals. The awards also serve to increase general awareness of green roofs, green walls, and other forms of living architecture while promoting their associated public and private benefits.

Appendix B:
Plant Lists

The following plant lists are specific to the projects in this book. They should not be blindly reproduced in the field. Many of these projects are conducting research. Plant survival is highly dependent on the suitability of those plants to the green roof's site conditions. Plants must be carefully matched to the roof location for both macroclimatic and microclimatic conditions. A professional horticulturist, landscape architect or local nursery business with green roof experience should be consulted for this information and to help select plants that meet the client's needs. An excellent reference book is Ed and Lucie Snodgrass' *Green Roof Plants* or Nigel Dunnett and Noel Kinsbury's *Planting Green Roofs and Living Walls*. Summaries of academic plant research across North America are now available on the Green Roofs for Healthy Cities website (www.greenroofs.org) in the Green Roofs Tree of Knowledge.

Solaire Building (Battery Park City, New York, New York)

Lotus coriculatus, Sagina subulata, Sedum requienii; Thymus Serphyllum; Miscanthus sinensis 'Gracillimus'; Pennesitum alopecuroides 'Hameln'; Antennaria dioica; Delosperma cooperi; Geranium sanguineum 'Johnson Blue'; Perovskia atriplicfolia; Sedum spurium 'Autumn Joy'; Cotoneaster dammeri 'Royal Beauty', Euonymus fortunei var. coloratus; Euonymus kiautsehovicas 'Manhattan'; Hypericum x 'Hidcote'; Juniper procumbens 'Nana'; Phyllostachys bissetti dwarf; Pinus mugo 'Mops'; Rosa Flower Carpet series; Spirea japonica 'Little Princess'; Phyllostachys Bissetti dwarf.

North Beach Place (San Francisco, California)

Acer rubrum 'Red Sunset' (Red Sunset Maple); *Acer circinatum* (Vine Maple); *Arbutus* X 'Marina' (Marina Strawberry Tree); *Ginkgo biloba* 'Fairmont' (Fairmont Ginkgo); *Malus floribundus* 'Harvest Gold' (Harvest Gold Crabapple); *Malus* X 'Red Jewel' (Red Jewel Crabapple); *Pittosporum undulatum* (Victorian Box); *Podocarpus gracilior* (African Fern Pine); *Podocarpus macrophyllus* (Buddhist Pine); *Tristania laurina* (Small Leaved Tristania); *Ulmus parvifolia* (Chinese Elm); *Zelkova serrata* 'Green Vase' (Green Vase Zelkova); *Abelia* X *grandiflora* 'Prostrata' (Dwarf Abelia); *Agapanthus* 'Peter Pan'; *Agapanthus* 'Tinkerbell' (Lily Of The Nile); *Asparagus densiflorus* 'Sprengeri' (Asparagus Fern); *Asplenium bulbiferum* (Mother Fern); *Aucuba japonica* 'Nana' (Dwarf Japanese Aucuba); *Choisya ternata (*Mexican Orange Blossom); *Cistus skanbergi* (Rock Rose); *Clivia miniata* (Kaffir Lily); *Coleonema pulchrum* (Breath Of Heaven); *Coprosma* 'Coppershine' (Coppershine Coprosma); *Coprosma repens* 'Marble Queen' (Marble Queen Coprosma); *Correa* 'Dusky Bells' (Dusky Bells Correa); *Dicksonia antarctica* (Tasmanian Tree Fern); *Dietes vegeta* (Fortnight Lily); *Euryops pectinatus* 'Viridis' (Green Leaved Euryops); *Hebe* 'Patty's Purple' (Patty's Purple Hebe); *Hemerocallis* 'Stella De Oro'; *Hemerocallis* 'Lemon Drop' (Hybrid Daylily); *Kniphofia* 'Bressingham Comet' (Dwarf Torch Lily); *Kniphofia* 'Little Maid' (Dwarf Yellow Torch Lily); *Lantana* 'Patriot Cowboy' (Patriot Cowboy Lantana); *Lavandula angustifolia* 'Hidcote' (English Lavender); *Lavandula stoechas* 'Otto Quast' (Spanish Lavender); *Leucothoe axillaries* (Coast Leucothoe); *Mahonia aquifolium* 'Golden Abundance' (Golden Abundance Oregon Grape); *Myrtus communis* 'Compacta' (Dwarf Myrtle); *Nandina domestica* 'Nana' (Dwarf Heavenly Bamboo); *Osmanthus delavayi* (Delavay Osmanthus); *Pennisetum setaceum* (Fountain Grass); *Philodendron* X 'Xanadu' (Xanadu Philodendron); *Phormium cookianum* (Mountain Flax); *Pittosporum tobira* 'Wheeler's Dwarf' (Wheeler's Dwarf Pittosporum); *Polystichum munitum* (Sword Fern); *Punica granatum* 'Nana' (Dwarf Pomegranate); *Rhododendron* X 'Cristo Rey' (Cristo Rey Rhododendron); *Rhododendron* 'Snow Lady' (Snow Lady Rhododendron); *Rosmarinus* 'Huntington Carpet' (Creeping Rosemary); *Ruscus hypoglossum* (Butcher's Broom); *Sarcoccoca hookerana humilis* (Sweet Box); *Stipa tenuissima* (Feather Grass); *Woodwardia fimbriata* (Chain Fern); *Aspidistra eliator* (Cast Iron Plant); *Chamaemelum nobile* 'Trenague' (Lawn Chamomile); *Convolvulus mauritanicus* (Ground Morning Glory); *Cotoneaster* 'Lowfast' (Lowfast Cotoneaster); *Dymondia margaretae* (Silver Carpet); *Heuchera* 'Bressingham Hybrids' (Coral Bells); *Iberis sempervirens* (Evergreen Candytuft); *Lamium maculatum* (Dead Nettle); Liriope 'Big Blue'; Liriope 'Silver Dragon', Plectranthus ciliatus (Spur Flower); *Polygonum capitatum* (Knotweed); *Sutera cordata* (Bacopa); *Campsis grandiflora* 'Morning Calm' (Chinese Trumpet Creeper); *Clematis armandii* (Evergreen Clematis); *Clytostoma callistegioides* (Violet Trumpet Vine); *Hardenbergia violacaea* (Lilac Vine); *Wisteria sinensis* (Chinese Wisteria).

Eastern Village Cohousing Condominiums (Silver Spring, Maryland)

The primary sedums used include *S. kamtschaticum, S. sexangulare, S. spurium Fuldaglut,* and *S. Weinenstaphaner Gold. Allium schoenoprasum, Orostachys aggregatum,* and *Talinum calycinum* complement the Sedum.

10th @ Hoyt Apartments
(Portland, Oregon)

Acer palmatum var. (Japanese Maple); *Magnolia virginiana* (Sweet Bay Magnolia); *Taxus baccata* (Irish Yew); *Ligustrum japonicum* (Japanese Privet); *Ilex crenata convexa* (Japanese Holly); *Mahonia nervosa* (Creeping Oregon Grape); *Camellia sasanqua* 'Yule Tide'; *Gunnera tinctoria* (Gunnera); *Pachysandra terminalis* (Japanese Spurge); *Galium odoratum* (Sweet Woodruff); *Ilex crenata* 'Sky Pencil'; *Buxus sempervirons suffruticosa* (True Dwarf Boxwood); *Helichrysum petiolare* (White Licorice); *Cupressus sempervirens* 'Tiny Tower' (Tiny Tower Italian Cypress); *Bacopa*.

Lot 8 Santa Lucia Preserve
(Carmel, California)

Extensive
Elymus arenarius; Euphorbia mysinites; Euphorbia polychrome; Festuca rubra; Festuca glauca 'Meerblau'; *Geum triflorum; Leucojum aestivum; Muscari armeniacum; Origanum* 'Kent Beauty'; *Ranunculus acris; Trifolium pretense; Sedum acre; Sedum album; Sedum integrifolium; Sedum oreganum; Sedum saxangulare; Sedum spurium;* 'Dragon's Blood'; *Sedum reflexum.*

Intensive
Allium cernum; Aster novae-angelica 'Purple Dome'; *Aster novae-angelica* 'Honeysong Pink'; *Centranthus ruber* 'Coccineus'; *Calamagrostris x a.* 'Karl Foerster'; *Carex morrowii* 'Variegata'; *Carex pendula; Eupatorium purpureum; Euphorbia amigdaloides; Euphorbia characias wulfenii; Geranium x oxonianum* 'Claridge Druce'; *Hemerocallis* 'Flava'; *Hosta plantagiana* 'Honey Bells'; *Hydrangea macrophylla; Liatris spicata; Lonicera nitida* 'Baggesen's Gold'; *Lonicera sempervirens; Miscanthus sinensis* 'Gracillimus'; *Nandina domestica* 'Plum Passion'; *Panicum virgatum* 'Strictum'; *Phlomis russelliana; Polystichum munitum; Pieris taiwanensis* 'Snow Drift'; *Ruta graveolens* 'Jackmans Blue'; *Sarcococca hookeriana* 'Digyma'; *Solidago canadensis* 'Baby Gold'; *Spiraea japonica* 'Little Princess'; *Tradescantia virginiana; Weigela florida* 'Minuet'.

The Louisa (Portland, Oregon)

Extensive Roof
Elymus arenarius; uphorbia mysinites; uphorbia polychrome; Festuca rubra; Festuca glauca 'Meerblau'; *Geum triflorum; Leucojum aestivum; Muscari armeniacum; Origanum* 'Kent Beauty'; *Ranunculus acris; Trifolium pretense; Sedum acre; Sedum album; Sedum integrifolium; Sedum oreganum; Sedum saxangulare; Sedum spurium* 'Dragon's Blood'; *Sedum reflexum.*

Intensive Roof
Allium cernum; Aster novae-angelica 'Purple Dome'; *Aster novae-angelica* 'Honeysong Pink'; *Centranthus ruber* 'Coccineus'; *Calamagrostris x a.* 'Karl Foerster'; *Carex morrowii* 'Variegata'; *Carex pendula; upatorium purpureum; Uphorbia amigdaloides; Uphorbia characias wulfenii; Geranium x oxonianum* 'Claridge Druce'; *Hemerocallis* 'Flava'; *Hosta plantagiana* 'Honey Bells'; *Hydrangea macrophylla; Liatris spicata; Lonicera nitida* 'Baggesen's Gold'; *Lonicera sempervirens; Miscanthus sinensis* 'Gracillimus'; *Nandina domestica* 'Plum Passion'; *Panicum virgatum* 'Strictum'; *Phlomis russelliana; Polystichum munitum; Pieris taiwanensis* 'Snow Drift'; *Ruta graveolens;* 'Jackmans Blue'; *Sarcococca hookeriana* 'Digyma'; *Solidago canadensis* 'Baby Gold'; *Spiraea japonica* 'Little Princess'; *Tradescantia virginiana; Weigela florida* 'Minuet'.

Ducks Unlimited National Headquarters & Oak Hammock Marsh Interpretive Centre
(Oak Hammock Marsh, Manitoba)

Blue Grama; Crocus; Three Flowered Aven; Little Bluestem; Long-headed Coneflower; Alumroot; Narrow-leaved; Sunflower; Purple Prairie Clover; arly **(*Early? Thanks.)** Blue Violet; Heart-leaved Alexander; Purple Prairie Violet; Black-eyed Susan; Western Wheatgrass; White Upland Aster; Wild Bergamot; Western Silvery Aster; Pink Flowered Onion; Meadow Blazingstar; Harebell; Northern bedstraw.

The above does not include the prairie grass mixture that is currently the private information of Ducks Unlimited Canada, who research and maintain an inventory of prairie grass seeds.

Peggy Notebaert Nature Museum
(Chicago, Illinois)

Extensive
Achillea millefolium 'Heidi' ('Heidi' yarrow); *Achillea sp.* ('Schwellenburg' yarrow); *Allium canadense* (wild onion); *Allium cernuum* (nodding wild onion); *Amorpha canescens* (leadplant); *Andropogon scoparius* (little bluestem grass); *Anemone patens wolfgangiana* (pasque flower); *Aquilegia canadensis* (American columbine); *Asclepias tuberosa* (butterfly weed); *Asclepias verticillata* (whorled milkweed); *Aster azureus* (sky blue aster); *Aster laevis* (smooth blue aster); *Aster ptarmicoides* (upland white aster); *Bouteloua curtipendula* (side-oats grama); *Buchloe dactyloides* (buffalo grass); *Campanula rotundifolia* (harebell); *Carex bicknellii* (Bicknell's sedge); *Coreopsis palmata* (prairie coreopsis); *Danthonia spicata* (poverty oat grass); *Dianthus allwoodii* ('Helen' carnation); *Dianthus gratianopolitanus* ('Spotty' carnation); *Dodecatheon meadii* (shooting star); *Geum triflorum* (prairie smoke); *Helianthus mollis* (downy sunflower); *Helianthus occidentalis* (western sunflower); *Heuchera richardsonii* (prairie alum root); *Koeleria cristata* (June grass); *Lavandula angustifolia* 'Hidecote' (Hidecote lavender); *Liatris aspera* (rough blazing star); *Parthenium integrifolium* (wild quinine); *Petalostemum candidum* (white prairie clover); *Petalostemum purpureum* (purple prairie clover); *Phlox bifida* (sand phlox); *Phlox pilosa* (downy phlox); *Sedum acre* (wall pepper); *Sedum album* (white sedum); *Sedum kamtschaticum* (orange stonecrop); *Sedum* 'Mochren' (Mochren sedum); *Sedum spurium* (sedum); *Sedum* 'Vera Jameson' (Vera Jameson sedum); *Sempervivum arachnoideum* (hens and chicks); *Solidago speciosa* (showy goldenrod); *Sporobolus heterolepis* (prairie dropseed); *Stachys byzantina* ('Helene von Stein'); *Thymus serpyllum* (thymus); *Celastrus scandens* (American bittersweet); *Clematis virginiana* (virgins bower).

Intensive
Aster sagitarius drumondii (Drummond's aster); *Blephilia ciliata* (Ohio horse mint); *Carex gravida* (sedge); *Carex pennsylvanica* (pennsylvania sedge); *Echinacea purpurea* (purple coneflower); *Elymus villosus* (silky wild rye); *Geranium sanguineum* 'Max Frei' (cranesbill); *Geranium sanguineum var.; striatum* (cranesbill); *Hemerocallis* ('Little Wine Cup' daylily); *Hystrix patula* (bottlebrush grass); *Parthenium integrifolium* (wild quinine); *Penstemon pallidus* (pale beard

tongue); *Polemonium reptans* (Jacobs ladder); *Smilacina racemosa* (false Solomons seal); *Solidago flexicaulis* (broad-leaved goldenrod); *Tradescantia ohiensis* (common spiderwort); *Rhus aromatica* 'Gro-low' (gro-low sumac); *Quercus imbricaria* (shingle oak).

Wetland

Acorus calamus (sweet flag); *Alisma subcordatum* (common water plantain); *Asclepias incarnata* (swamp milkweed); *Caltha palustris* (marsh marigold); *arex cristatella* (crested oval sedge); *Carex lacustris* (common lake sedge); *Equisetum arvense* (horsetail); *Eupatorium maculatum* (spotted Joe Pye weed); *Helenium autumnale* (autumn sneezeweed); *Iris virginica shrevei* (blue flag iris); *Juncus dudleyi* (Dudley's rush); *Juncus; ffusus* (common rush); *Juncus torreyi* (Torry's Rush); *Lobelia cardinalis* (cardinal flower); *Lobelia siphilitica* (great blue lobelia); *Panicum virgatum* (switch grass); *Pontedaria cordata* (pickerel weed); *Sagittaria latifolia* (common arrowhead); *Scirpus atrovirens* (dark green rush); *Scutellaria; pilobiifolia* (marsh skullcap); *Solidago riddellii* (Riddel's Goldenrod); *Sparganium; urycarpum* (bur-reed); *Spartina pectinata* (prairie cordgrass); *Spiraea alba* (meadowsweet); *Verbena hastata* (blue verbena); *Vernonia fasciculata* (Ironweed); *Veronicastrum virginicum* (Culver's root); *Zizia aurea* (golden Alexanders).

The Church of Jesus Christ of Latter-Saints Convention Center (Salt Lake City, Utah)

Quercus robur (English Oak); *Tilia Americana* "Redmond" (Redmound American Linden); *Ulmus parvifolia* (Lacebark Elm); *Zelkova serrata* (Japanese Zelkova); *Populus angustilolia* (Narrowleaf Aspen); *Quercus macrocarpa* (Bur Oak); *Acer griseum* (Paperbark Maple); *Acer glabrum* (Rocky Mountain Maple); *Acer grandidentatum* (Big Tooth Maple); *Gymnocladus dioica* (Kentucky Coffee Tree); *Quercus gambelli* (Gambels Oak); *Sophora japonica* (Japanese Scholar Tree); *Pinus leucodemis* "Heldrechii" (Bosnian Pine); *Pinus aristata* (Bristlecomb Pine); *Pinus flexilis* 'Vanderwolfe' (Vanderwolfe Limber Pine); *Picea omorika* (Serbian Spruce); *Pinus jeffreyi* (Jeffrey Pine); *Pseudotsuga menziesii* (Glauca Douglas Fir); *Pinus sylvestris* (Scoth Pine); *Acer buergeranum* (Tridant Maple); *Amelanchier Canadensis* (Serviceberry); *Aesculus pavia* (Red Buckeye); *Amelanchier Saskatoon* (Saskatoon Amelanchier); *Betula fontinalis* (Western Birch); *Cercocarpus ledifolius* (Mountain Mahogany); *Ceris occidentalis* (Western Redbud); *Sorbus aucuparia* (Mountain Ash); *Aronia arbutifolia* (Red Chokeberry); *Alnus tenuifolia* (Mountain Adler); *Rhus typhina* (Staghorn Sumac); *Viburnum dentatum* (Arrowwood Viburnum); *Viburnum trilobum* (Highbush Cranberry); *Aronia melanocarpa* (Black Chokeberry); *Cornus sericea* (Red Osier Dogwood); *Hydrangea quercifolia* 'Snow Queen' (Oakleaf Hydrangea); *Juniperus chinense* 'Plitzeriana' (Plitzer Juniper); *Viburnum chinensis* 'Compactum' (Compact Fragrant Viburnum); *Viburnum opulus* 'Compactum' (Compact Cranberry Viburnum); *Asarum canadense* (Wild Ginger); *Cotoneaster salicifolia* 'Repandens' (Willowleaf Cotoneaster); *Chasmanthus latitolia* (Northern Seed Oats); *Juniperus horizontalis* (Creeping Juniper); *Lysimachia hummularia* (Moneywort); *Mahonia repens* (Creeping Mahonia); *Ribes aureum* (Golden Currant); *Rosa* 'Bonica' "Improved" (Bonica Rose); *Rosa* 'Pink Meidiland' (Pink Meidiland Rose); *Rosa* 'Scarlet Meidiland' (Scarlet Meidiland Rose); *Rosa* 'The Fairy' (The Fairy Rose); *Symphoricarpus oreophilus* (Mountain Snowberry); *Vinca minor* (Periwinkle); *Aquilegia caerulea* (Blue Columbine); *Aquilegia chrysantha* (Yellow Columbine); *Baptisia auslralis* (Blue Baptisia); *Castilleja chromosa* (Early Indian Paintbrush); *Centaurea cyanus* (Bachelor Buttons); *Cheiranthus allionii* (Wallflower); *Delphinium ajacis* (Rocket Larkspur); *Thalictrum dasycarpum* (Meadowrue); *Filipendula venusia* 'rubra' (Meadowsweet); *Geranium maculatum* 'Chatto' (Wild Geranium); *Geranium richardsonii* (Wild Geranium); *Gilia capitata* (Globe Gilia); *Gilia leptantha* 'purpusil' (Blue Gilia); *Lupinus* 'Garden Gnome' (Garden Gnome Lupine); *Lupinus perennis* (Wild Lupine); *Lupinus sericeus* (Silky Lupine); *Penstemon eatonli* (Bearded Tongue); *Penstemon rilldus* (Large-flower Penstemon); *Phiox longifolia* (Smooth Blue Beardtongue); *Pulmonaria longifolia* 'cerennenis' (Longleaf Lungwort); *Pulmonaria rubra* ('Redstart' Lungwort); *Thermopsis divaricarpa* (Foothills Golden Banner); *Asciepias tuberosa* (Butterflyweed); *Artemisia ludoviciana* (Common Sage); *Artemisa frigida* (Pasture Sage); *Aster englemanii* (Engleman Aster); *Campanula rolundifolia* (Harebell); *Castilleja linanaefolia* (Wyoming Paint Brush); *Coreopsis lanceolata* (Lanceleaf Coreopsis); *Coreopsis linotoria* (Coreopsis); *Cosmos bipinnatus* (Cosmos); *Eupatorium maculatum* (Joe Pye Weed); *Echinacea purpurea* (Purple coneflower); *Erigeron speciosus* (Fleabane); *Gaillardia aristata* (Blacketflower); *Gaillardia putchella* (Fireweed); *Geranium x cantabrigiense* (Cambridge Cranesbill); *Geranium himalayense* 'Plenum' (Dot Lavender Cranesbill); *Geranium sanguineum* (Bloody Cranesbill); *Geranium viscosissimum* (Sticky Geranium); *Helianthus maximiliani* (Maximillian Sunflower); *Helianthus annuus* (Sunflower); *Heliopsis helianthoides* 'Summer Sun' (OxEye Sunflower); *Iris missouriensis* (Blue Iris); *Liatris pycnostachya* (Gayfeather); *Liatris spicata* (Dense Blazingstar); *Linaria maroccana* (Baby Snapdragon); *Penstemom cyananthus* (Bearded Tongue); *Ratibida columnaris* (Prarie Coneflower); *Silene amenia* (Sweet William); *Verbena stricta* (Hoary Vervain); *Veronicastrum virginicum* (Culvers Root); *Aster novae-angliae* 'Alma Potschke' (New England Aster); *Aster azureus* (Sky Blue Aster); *Aster laevis* 'Blue Bird' (Smooth Aster); *Engeron speciosus* (Aspen Daisy); *Helianthus mollis* (Downey Sunflower); *Solidago rigida* (Meadow Goldenrod); *Solidago* 'Golden Baby' (Golden Baby Goldenrod); *Agropyron spicatum* (Bluebunch Wheatgrass); *Chasmanthus latifolia* (Northern Seed Oats); *Dactylis glomerata* (Orchard Grass); *Festuca idahoensis* (Bluebunch Fescue); *Poa sandebergii* (Sandberg Blue Grass); *Sporobolus cruptandrus* (Sand Dropseed); *Stipa comata* (Needle and Thread Grass); *Triticum x Elytrigia* (Regreen Grass); *Parthenocissus quinquefolia* (Virginia Creeper); *Camassia soilloides* (Wild Hyacinth); *Crocus* 'Snow Bunting' (White Crocus); *Crocus pulchellus* (Crocus); *Narcissus* 'Jack Snipe' (Dwarf Daffodil); *Narcissus* 'Tete a Tete' (Dwarf Daffodi); *Muscari azureum* (Grape Hyacinth); *Scilla campanulata Alba* (Wood Hyacinth).

Oaklyn Branch, Evansville Vanderburgh Public Library (Evansville, Indiana)

Andropogon scoparious; *Bouteloua* curitpendula; *Centaurea* cyanus; *Campanula* rotundifolia; *Carex* annectans; *Carex* bicknellii; *Coreopsis* tinctoria; *Elymus* Canadensis; *Liatris* spicata; *Phlox* drummondii; *Phlox* pilosa; *Sphaeralcea* coccinea; *Sporobolus* heterolepis.

Life Expressions Wellness Center

(Sugar Loaf, Pennsylvania)

Allium schoenprasm; *Dianthus* deltoids; *Sedum* acre; *Sedum* album (Coral Carpet); *Sedum* floriferum; *Sedum* origanum; *Sedum* reflexum; *Sedum* sarmentosum; *Sedum* sexangulare; *Sedum* spurium Fuldiglut; *Sedum* spurium Tricolor.

Ballard Branch of the Seattle Public Library

(Seattle, Washington)

Achillea tomentosa (Woolly yarrow); *Armeria maritima* (Sea pink, sea thrift); *Carex inops pensylvanica* (Long-stoloned sedge); *Eriphyllum lanatum* (Oregon sunshine); *Festuca rubra* (Red creeping fescue); *Festuca idahoensis* (Idaho fescue); *Phlox subulata* (Creeping phlox); *Saxifrage cespitosa* (Tufted saxifrage); *Sedum oreganum* (Oregon stonecrop); *Sedum album* (White stonecrop); *Sedum spurium* (Two-row stonecrop); *Sisyrinchium idahoensis* (Blue-eyed grass); *Thymus serphyllum* (Thyme); *Triteleia hyacintha* (Fool's onion).

Nashville Public Square

(Nashville, Tennessee)

Acer rubrum 'October Glory' (Red Maple); *Acer saccharum 'Legacy'* (Sugar Maple); *Cercis Canadensis* (Redbud); *Cornus Kousa* (Dogwood); *Crataegus phaenopryum* (Washington Hawthorne); *Gleditsia triacanthos inermis 'halka'* (Honey Locust); *Ilex opaca* (American Holly); *Liquidambar styraciflua 'Rotundiloba'* (Sweetgum); *Magnolia grandiflora 'Claudia Wannamaker'*; (Southern Magnolia); *Nyssa sylvatica 'Forum'* (Black Gum); *Quercus alba* (White Oak); *Quercus rubra* (Northern Red Oak); *Quercus lyrata* (Overcup Oak); *Quercus muehlengergii* (Chinkapin Oak); *Quercus nuttallii* (Nuttall Oak); *Quercus shumardii* (Southern Red Oak); *Ulmus americana 'Princeton'* (American Elm); *Berberis x gladwynensis 'William Penn'* (Barberry); *Clethra alnifolia 'Hummingbird'* (Dwarf Clethra); *Fothergilla gardenia* (Dwarf Fothergilla); *Hydrangea quercifolia* (Oakleaf Hydrangea); *Hypericum calycinum* (Hypericum); *Ilex opaca 'Maryland Dwarf'* (Dwarf American Holly); *Ilex verticillata 'Winter Red'* (Deciduous Holly); *Ilex glabra 'Densa'* (Dwarf Inkberry); *Itea virginica 'Henrys Garnet'* (Itea); *Itea virginica 'Little Henry'* (Dwarf Itea); *Prunus laurocerasus 'Otto Luyken'* (O. L. Laurel); *Rosa laevigata* (Cherokee Rose); *Panicum virgatum 'Prairie Sky'* (Switchgrass); *Sorghastrum nutans 'Sioux Blue'* (Indian Grass); *Astilbe x arendsii 'Red Sentinel'* (Astilbe); *Baptista australis* (False Blue Indigo); *Coreopsis lanceolata 'Grandiflora'* (Grand Coreopsis); *Dryopteris marginallis* (Marginal Wood Fern); *Echinacea purpurea 'Bright Star'* (Purple Coneflower); *Gelsemium sempervirens* (Yellow Jessamine); *Hemerocallis x 'Stella d'Oro'* (Daylily); *Hosta plantaginea* (Fragrant Plantain-Lily); *Liatrus spicata* (Gayfeather); *Osmunda cinnamonea* (Cinnamon Fern); *Polygonatum biflorum* (Solomon's Seal); *Rudbeckia fulgida 'Goldstrum'* (Black-eyed Susan).

Ford Rouge Dearborn Truck Plant

(Dearborn, Michigan)

Sedum 'Fulda Glow' (cuttings); *Sedum 'Diffusum'* (cuttings); *Sedum acr; Sedum kamtschaticum; Sedum ellacombeanum; Sedum album ; Sedum pulchellum; Sedum 'Coccineum'; Sedum reflexum.*

Heinz 57 Center

(Pittsburgh, Pennsylvania)

The plants range from low-growing groundcovers like *Sedum spurium* and *Phlox pilosa* to tall varieties like *Carex annectens, Anthemis tinctoria, Dianthus deltoides, Festuca ovina* and *Chrysanthemum leucanthemum.*

601 Congress Street, Seaport District

(Boston, Massachusetts)

Bouteloua cortipendula (Side Oats Grama); *Seslaria autumnalis* (Autumn Moor Grass); *Sporabolus heteralepis* (Prairie Dropseed); *Miscanthus sinensis 'Graziella'* (Japanese Silver Grass); *Sedum kamtschaticum* (Kamschatka Stonecrop); *Sedum album* (White Stonecrop); *Sedum spurium 'Fuldaglut'* (Fuldaglut Two Row Stonecrop).

ABN AMRO Plaza

(Chicago, Illinois)

Acer freemanii 'Autumn Blaze', ginkgo biloba, crataegus phanopyrum, pinus sylvestris; cornus sericea 'Isanti', myrica pennsylvanica; calamogrostis 'Karl Foerster', eunonymus fortunei 'Coloratus', nepeta faasenii, perovskia atriplicifolia, veronica spicata 'Icicle'. Various annuals and seasonal plantings have also been installed including summer annuals, spring bulbs, and fall mums.

Calamos Investment

(Naperville, Illinois)

Sedum acre; Sedum kamtschaticum; Sedum kamtschaticum ellacombianum; Sedum reflexum; Sedum spurium; Sedum spurium coccineum; Sedum spurium 'Summer Glory'; Sedum sexangulare; Sedum album; Sedum rupestre; Sedum 'Fulda Glut'; Sedum middendorfianum 'Diffusum'.

Appendix C:
Standards

Green roof standards can be performance based or prescriptive. Performance based standards differ from prescriptive ones because they lay out measurable criteria that have to be fulfilled without determining how they are met; whereas prescriptive standards lay out guidelines which prioritize certain methods of achieving a desired goal. For example, in The City of Portland green roofs must be built to retain a specified amount of storm water runoff and attain the performance requirements of the city's storm water management strategy. Green roofs in Toronto must satisfy a list of prescribed requirements regarding size, depth, and vegetation in order to qualify for a green roof financial incentive. Standards can be issued by a municipality or an independent standard setting body.

Currently there are three non-governmental organizations (NGO) that have published material relating to green roof construction: ASTM International (ASTM), *German Forschungsgesellschaft Landschaftsentwicklung Landschafts bau e.V,* (FLL), and Factory Mutual Global (FM Global).

ASTM has published four *Green Roof Performance Standards* and one standard guide. Four more standards are under development.

    E 2396-05 Saturated Water Permeability of Granular Drainage Media.

    E2397-05 Determination of Dead Loads and Live Loads associated with Green Roof Systems.

    E2398-05 Water Capture and Media Retention Standards of Geocomposite Drain Layers for Green Roof Systems.

    E2399-05 Maximum Media Density for Dead Load Analysis of Green Roof Systems.

    E2400-06 Standard Guide for Selection, Installation, and Maintenance of Plants for Green Roofs.

In Germany, green roofs must meet certain standards regarding their quality and method of construction. In 1995, the *Forschungsgesellschaft Landschaftsentwicklung Landschafts bau e.V,* (FLL), the German landscaping industries' non-profit research and standard setting body, published the *Guidelines for the Planning, Execution, and Upkeep of Green Roof Sites* in English. These FLL documents, revised in 2002, are a set of guiding principles used by designers to determine which green roof systems are best suited to different buildings and climates in Germany.

The FLL Guidelines are prescriptive and specify elements like seam overlaps and vegetation-free zones. Many North American companies use FLL as a mark of their products' and systems' excellence. However, when drawing on the recommendations therein it is important to remember that these guidelines were produced for the German market and as such cannot always be directly applied in a North American context. Local Building Codes, ordinances, and other geographically specific legislation should take precedence over imported standards.

Though it is not a standard setting body, FM Global, a commercial and industrial property insurance and risk management organization, exerts a large amount of influence over the construction industry. It published *Property Loss Prevention Data Sheet 1-35: Green Roofs* in early 2006. This prescriptive document is largely based on FLL Guidelines and draws on other FM Data Sheets to establish standards for an *FM Approved Green Roof Assembly*.

Green Roofs for Healthy Cities is actively working with the Single Ply Roofing Industry (SPRI) to develop American National Standards Institute (ANSI) standards that when complete will provide design guidelines associated with wind uplift and fire prevention.

In North America, at least a dozen cities (e.g. Chicago, Portland, Washington, Minneapolis, and Toronto) have established programs that have, or will lead to, more widespread green roof implementation. Municipalities have taken both approaches but none of the current standards, with the exception of Chicago's Green Building/Green Roof Matrix, have been legally integrated into a Building Code. When designing and installing green roofs they should be built to local codes and ordinances under the guidance of existing guidelines and standards

Appendix D:
Green Roofs and Leadership in Energy Efficiency Design (LEED®)

As demonstrated by many of the award winning projects, green roofs can have a powerful positive impact on the sustainability of buildings. This impact is recognized within the U.S. Green Building Council's LEED® green building rating systems. The following is a summary of how green roofs can facilitate a significant improvement in the LEED® rating of a building in some cases securing or contributing as many as fifteen credits under the system. In some instances, green roofs may *contribute* to earning LEED® credits when used with other sustainable building elements.

PART 1: SUSTAINABLE SITES

Sustainable site development reduces the impact of the building and its site on the larger urban and global environment. Opportunities for a building that incorporates a green roof to reduce its impact on the surrounding environment include:

Reduced Site Disturbance, Protect or Restore Open Space – conserves existing areas and restores damaged areas to provide habitat and protect biodiversity. Earning this credit under the US LEED-NC is possible through the creation of suitable, restored habitat that could include native plant restoration and the creation of insect and bird habitat on a green roof if 50 per cent or more of the open space of the site is covered with a green roof of suitable, restored habitat.

Reduced Site Disturbance, Development Footprint Credit – While no credit has yet been awarded by the US Green Building Council to a green roof for fulfilling the Reduced Site Disturbance, Development Footprint Credit, it is conceivable that a green roof could fulfill the intent of this credit by "restoring damaged areas to provide habitat and promote biodiversity." Earning this credit may be possible in areas with no local zoning requirements for open space. In this case LEED® requires that the building "designate open space area adjacent to the building that is equal to the development footprint." If a building were to cover the entire site, as is often the case in dense urban environments, and installed a green roof that covered the entire building and provided a restored habitat it could fulfill the intent of this credit. *(See Oak Hammock Marsh or Feldman Residence)*

Landscape Design That Reduces Urban Heat Islands, Roof – aims to limit the amount of heat absorbed by the hard, dark surfaces of the urban environment such as roofing material, concrete, and asphalt that raise the temperature of a city as a whole and increase the energy consumed for cooling. Green roofs reduce the urban heat island by absorbing a significant percentage of the solar energy that reaches that roof surface and helps to moderate its temperature and that of the city as a whole. Green roofs that cover 50 per cent of the roof surface or 75 per cent of the total roof surface in combination with a "cool roof" earn a credit under LEED®.

PART 2: WATER EFFICIENCY

Storm water management and water conservation measures have the potential to save enormous amounts of water while reducing the demands on the urban infrastructure to supply water and treat waste water generated from a building and reduce the impact of large areas of hard surfaces on local water bodies. Green roofs are important elements in storm water management in that they can significantly reduce the amount and intensity of storm water discharged from the site. In many cases they have been effectively coupled with storm water collection systems to reduce or eliminate potable water use for irrigation and could be used to service a part of the building's water demands that are typically supplied by potable water. *(See Nashville Public Square or Evergreen State College)*

Storm Water Management – focuses on reducing the impacts of storm water flow from the site by reducing the rate and quantity of water discharged from the site and treating the storm water that does leave the site. Green roofs can function effectively in both roles, first by reducing or eliminating the rate and quantity of water discharged from the site through direct absorption and in many cases storage of the excess water. Intensive green roof systems with irrigation green roofs can be utilized to remove suspended sediments that supplement the roof's soil and nutrients absorbed by the roof's plants. A credit can be earned for both the reduction of the rate and quantity of storm water discharge and storm water treatment under the LEED® rating system.

Water Efficient Landscaping – focuses on reducing the quantity and improving the quality of storm water discharged from a site. A building can incorporate a collection system to store storm water from the buildings site and roof surfaces to be used for irrigation of the green roof and other landscape features. Two credits are earned if all potable water is eliminated for irrigation of the green roof and all other landscape features.

Water Use Reduction – one credit can be earned under this category for the reduction of potable water use in a building by 20 per cent and a second earned for a further 10 per cent reduction (30 per cent total) under LEED®. Projects that incorporate a storm water capture system can relatively easily and inexpensively expand the capacity of the storm water capture system to replace potable water with storm water for toilet flushing in some jurisdictions. *(See The Solaire)*

Innovative Wastewater Technologies – encourages the reduction of wastewater generation through water conservation and on-site treatment. Green roofs can be utilized as wastewater treatment media through a number of innovative techniques. Grey water treatment can be effectively accomplished by a green roof when carefully designed. Other features such as the incorporation of compost tea from a composting toilet is another innovative use of a green roof to aid in reducing wastewater generation. *(See Sanitation District N°1)*

PART 3: ENERGY AND ATMOSPHERE

The intent of this initiative is to realize opportunities to improve the performance of the building's envelope and increase the efficiency of the building systems. Green roofs can contribute to earning three or more Energy and Atmosphere Credits.

Optimise Energy Performance – focus on the opportunities to reduce energy consumption and realize the substantial opportunities for cost savings in building operation. Green roofs enhance the energy performance of buildings in a variety of ways. The growing medium and plants provide considerable thermal insulation and evapotransipiration thereby reducing the building's energy consumption. This is especially important in summer when the plants of the green roof absorb and reflect a significant portion of the suns rays, substantially reducing the building's cooling demand. Green roofs can contribute to the achievement of one or more of the ten points available for this credit.

Renewable Energy – reduces the reliance on traditional sources of energy in favour of renewable sources. Some studies have found that photovoltaic cells installed near a green roof are enhanced in efficiency because of a reduction in ambient air temperatures through evapotranspiration. When photovoltaic cells become hot they loose efficiency. In some green roof designs a photovoltaic cell can also be coupled with the irrigation system to deliver irrigation water in direct proportion to the water demand of the roof. By delivering additional water through the green roof's irrigation system in the hot summer months for evaporative cooling, a building to a large degree can cool itself by "sweating." A building earns a credit for providing 5 per cent of its total energy demand by alternative forms of energy production.

CFC and Ozone Depleting Substance Reduction – encourages the use of refrigerants that do not contain CFC's or HCFC's that harm the earth's ozone layer. Using a green roof to offset most of a building's entire cooling load is possible in some applications. *(See Oak Hammock Marsh)* By replacing a costly mechanical system that contain CFC's, and other ozone depleting substances, with green roofs a building can essentially eliminate all CFC and HCFC equipment satisfying the requirements of an Energy and Atmosphere Credit and the Ozone depletion credit.

PART 4: MATERIALS AND RESOURCES

The extraction of resources and energy consumption related to the manufacturing process of many building materials has a significant impact on the natural environment. Impacts such as the production of wastes by the building's occupants also have significant environmental impacts that can to some degree be mitigated by a green roof. A green roof typically will not earn any Materials and Resources Credits but can help to satisfy the requirements of the Storage and Collection of Recyclables Prerequisite.

Storage and Collection of Recyclables – Beyond the impacts of the building materials themselves, the materials consumed within the building also have a significant potential impact on the environment. In some applications green roofs can function effectively in combination with a food and paper composting program to divert food waste from the landfill while enriching the soil of the roof. Other recycled materials may be incorporated into the growing medium, such as compost and various aggregates such as crushed brick.

PART 5: INDOOR ENVIRONMENTAL QUALITY

The recent past has demonstrated a wide range of concerns with the quality of the indoor environment that range from simple issues of thermal comfort to a range of sick building issues that threaten human health. Today in an effort to counteract these negative impacts a number of steps have been taken that include flexible mechanical system designs that respond to fluctuating tenant demands, and maximizing natural light penetration that can be enhanced with the inclusion of a green roof.

Carbon Dioxide Monitoring – carbon dioxide monitors allow a building's mechanical system to respond to the occupancy level of a particular room or building zone by supplying a volume of ventilation air directly related to a room's occupancy. When sensitively incorporated in the design, ventilation inlets can be placed on the roof above the green roof. The incoming ventilation air is then "oxygen enriched," providing a higher quality of ventilation air at a low volume reducing energy consumption. While a green roof will not directly earn a credit under LEED® for Carbon Dioxide Monitoring it demonstrates how sensitive design that incorporates a green roof can support the LEED® Certification Process.

PART 6: INNOVATION IN DESIGN

Although LEED® provides many opportunities to earn credits for the incorporation of a green roof in a building design, it also provides many opportunities to innovate beyond the set credits outlined within the five categories of the rating system. Innovation in Design credits allow a green roof designer to think outside the box and create innovative sustainable solutions to building challenges and earn additional credit under the system.

Appendix E:
Selected References and Resources

Conference Proceedings

Each year, Green Roofs for Healthy Cities calls for papers on green roof research, policy, and design and publishes the peer reviewed papers presented at its annual conference. These papers contain the most up to date information on green roof design and implementation, technical research, and policy developments. The only compilation of green roof data from around the world, these CDs are a must for anyone involved in green roof design, implementation, installation, research, policy development, and outreach activities.

Text and Articles

Dunnett, Nigel and Noel Kingsbury. (2004) *Planting Green Roofs and Living Walls*. Portland, OR: Timber Press.

Earth Pledge. (2005) *Green Roofs: Ecological Planning & Design*. New York: Schiffer Pub. Ltd.

Osmundson, Theodore. (1999) *Roof Gardens: History, Design, and Construction*. New York: W. W. Norton & Company, Inc.

Johnston, Jacklyn and John Newton. (1993) *Building Green: A Guide To Using Plants On Roofs, Walls And Pavements*. London, UK: The London Ecology Unit.

Peck, Steven and Monica Kuhn. (March 1999) *Greenbacks from Green Roofs: Forging a New Industry in Canada*. Ottawa, ON: Canadian Mortgage and Housing Corporation.

Robèrt, K-H, H. Daly, P. Hawken, and J. Holmberg. (1997) A Compass for Sustainable Development. *International Journal of Sustainable Development and World Ecology*, 4 (1997): 79-92.

Snodgrass, Edmund and L. Lucie. (2006) *Green Roof Plants: A Resource & Planting Guide*. Portland, OR: Timber Press.

Thomson, J. William and Kim Sorvig. (2000). *Sustainable Landscape Construction: A Guide to Green Building Outdoors*. Washington, DC: Island Press.

Trusty, Wayne B. and Scot Horst. (November 2002) *Integrating LCA Tools in Green Building Rating Systems*. The Austin Papers: Best of the 2002 International Green Building Conference.

Werthermann, Christian. (2007) *Green Roof – A Case Study: Michael Van Valkenburgh Associates' Design for the Headquarters of the American Society of Landscape Architects*. New York, NY: Princeton Architectural Press.

Online Resources

www.greenroofs.org – Green Roofs for Healthy Cities' official website. This website contains intensive information about green roofs, their installation, benefits, and demonstration projects. The resources available include upcoming training courses, all the back issues of the Green Roofs Infrastructure Monitor, other Green Roofs for Healthy Cities publications, useful websites, and other publications. Searhcable files of our Individual and Corporate Members are provided as well as services like the *GreenSave Calculator*, a Life Cycle Cost Benefit tool, and the *Green Roofs Tree of Knowledge*, a searchable database of research paper and policy initiative summaries.

www.greenroofs.com – Has the directory of Green Resources News, upcoming events, recommended readings, a project database, latest news about green roofs including projects and research links.

www.usgbc.org – Official website of United States Green Building Council. This website contains detailed information about LEED®, latest news and events, all the USGBC chapters' websites, and Industry publication and links.

www.cagbc.ca – Official website of Canadian Green Building Council. The website features the most recent green building projects, news, and upcoming events. All information about the building rating system and other resources are included in the website.